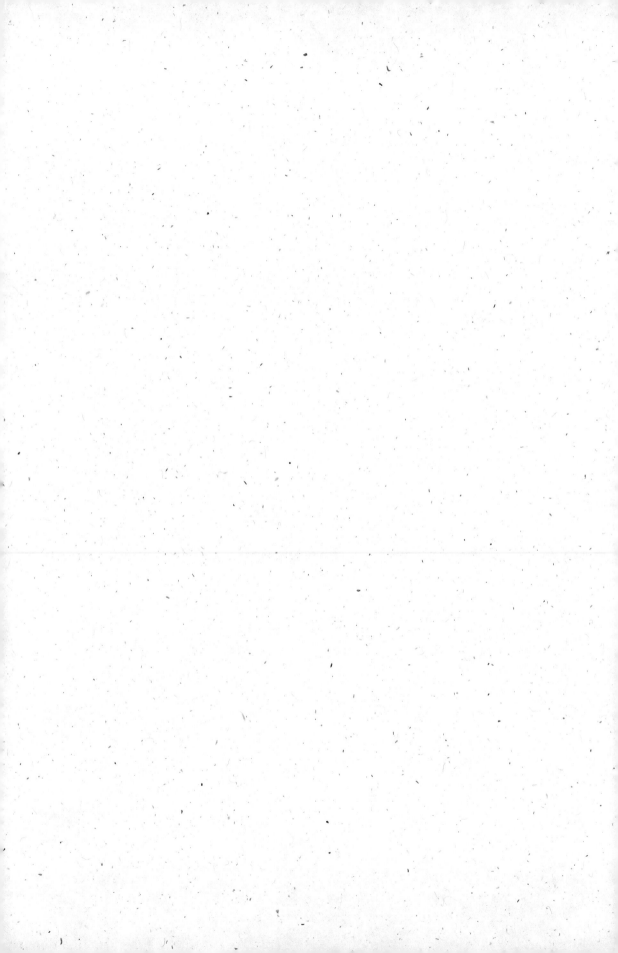

Sharks: *The Animal Answer Guide*

Sharks
The Animal Answer Guide

Gene Helfman and George H. Burgess

Johns Hopkins University Press Baltimore

© 2014 Johns Hopkins University Press
All rights reserved. Published 2014
Printed in the United States of America on acid-free paper
9 8 7 6 5 4 3 2 1

Johns Hopkins University Press
2715 North Charles Street
Baltimore, Maryland 21218-4363
www.press.jhu.edu

Library of Congress Cataloging-in-Publication Data

Helfman, Gene S.
 Sharks : the animal answer guide / Gene Helfman and George H. Burgess.
 pages cm
 Includes bibliographical references and index.
 ISBN 978-1-4214-1308-2 (hardcover : alk. paper) — ISBN 978-1-4214-1309-9
(pbk. : alk. paper) — ISBN 978-1-4214-1310-5 (electronic) — ISBN 1-4214-
1308-6 (hardcover : alk. paper) — ISBN 1-4214-1309-4 (pbk. : alk. paper) —
ISBN 1-4214-1310-8 (electronic) 1. Chondrichthyes—Miscellanea.
2. Sharks—Miscellanea. I. Burgess, George H., 1949– II. Title.
 QL638.6.H45 2014
 597.3—dc23 2013027724

A catalog record for this book is available from the British Library.

*Special discounts are available for bulk purchases of this book. For more information,
please contact Special Sales at 410-516-6936 or specialsales@press.jhu.edu.*

Johns Hopkins University Press uses environmentally friendly book materials,
including recycled text paper that is composed of at least 30 percent post-consumer
waste, whenever possible.

Contents

Acknowledgments

Knowledge about sharks and their relatives grows daily. Long before this information appears in the scientific literature, it is generously shared with other shark lovers and researchers. Many of the facts and ideas presented here fall into this category of new, verified information that has not yet quite become publicly accessible. In addition, much "information" about sharks appears regularly on websites, without reliably identifiable sources. Our colleagues around the world have aided us in finding the former and verifying the latter, for which we are exceedingly grateful.

For permission to use illustrations and for providing or verifying information on various subjects we thank a host of people, including Aaron Fisk, Adam Summers, Austin Gallagher, Bruce Carlson, Bruce Masse, Cam Macdonald, Carolyn Moser, Colin Simpfendorfer, Dave Ebert, Dave Hally, Dave Harasti, Dave Sumang, Dayv Lowry, David Sims, Dovi Kacev, Duncan Elkins, Ellis Loew, Greg Cailliet, Harry Fierstine, Jack Randall, Jane Watson, J. B. Heiser, Jean Spalti, Jeff Rotman, Jenny Kemper, Jim Whittington, Jimmy Bernot, Joe Gaydos, Johann Mourier, John Ford, John McCosker, Jose Castro, Judy Meyer (!!), Kara Yopak, Kaye DeHays, Ken Goldman, Kyle Mara, Laura Kozuch, Maria Pickering, Marj Awai, Mark Hixon, Matthew McDavitt, Michal Heithaus, Misty Paig-Tran, Mo Donnelly, Nick Dulvy, Olaf Weyl, Paul Skelton, Paul Vecsei, Phil Motta (!), Phillip Weemstra, Ralph Collier, Randall Dahn, Randy Honebrink, Ryan Kempster, Samantha Mulvany, Sonny Gruber, the South African Institute of Aquatic Biology (SAIAB), the Whale Interpretive Centre, Tim Berra, Tim Higham, Tom Emerson, Tom Reinert (!), Wes Pratt, and Zeb Hogan.

Our special thanks to Vincent Burke, Jennifer Malat, and Sara Cleary of Johns Hopkins University Press for their patience with and understanding of our halting progress in compiling this book. They proved that the publishing industry is far from the heartless monolith that authors often claim it to be.

And in particular we thank our wives, Judy Meyer and Linda Schulze Burgess, for their encouragement, patience, and crucial support, without which this book could never have been completed.

One historical note. Books about sharks abound. Many aim at an elementary level and a fairly young audience, are filled with photographs or drawings, sensationalize or overemphasize the spectacular or speculative, and provide little technical detail for someone interested in more than basics. At the other extreme are the scholarly textbooks and volumes written

by experts, targeted at a college or professional audience, providing great detail or the latest research findings for persons who already possess an extensive biological education. Comparatively few are written "for people with little prior knowledge of sharks, but with a curiosity and interest in finding out more about them" (Springer and Gold 1989, p. 10). Our intended readers are the same for which Springer and Gold wrote *Sharks in Question*, namely, intelligent people who want more than the basics without the hype and aren't afraid of the technical detail necessary to fully appreciate why sharks are worthy of study. Springer and Gold accomplished this superbly, and we hope to reach the same audience and level of detail, updating the material in that now out-of-print volume.

Springer and Gold concluded their influential 1989 volume by observing, "You can be certain that many more books about sharks will be written as new information increases our understanding; quite possibly, the future authors are among our readers" (p. 145). They were correct on both counts. We are grateful for their inspiration and hope that we too will stimulate future efforts at understanding, explaining, and promoting the natural history, biology, behavior, and conservation of these fascinating animals.

Introduction

From Gene Helfman

George Burgess and I both became fascinated with sharks at an early age. My shark "career" began when I was maybe 9 or 10. Someone landed a Basking Shark at the Santa Monica Pier in southern California. The *Los Angeles Times* ran an article with a photo, including an interview with a biologist from the California Department of Fish and Game, Dr. John Fitch. At my father's encouragement, I wrote Dr. Fitch a letter that I'm sure I would find embarrassing today; I probably asked some basic questions and said what a neat animal the shark was. To my astonishment, he wrote back. That's really all it took. I was hooked. Someone important thought my interest was worthwhile.

I read just about every shark book available in English before I was 12. All my reports, for science or other classes, were about sharks. When I entered Van Nuys High School, I hooked up with a couple of other shark nuts, and we formed a group we called the Shark Research Committee. (The Shark Research Committee still exists, still chaired by Ralph Collier, who has since written a book on shark attacks along the Pacific coast.) Encouraged by a biology teacher, we contacted Dr. Marshall Urist at UCLA, a researcher interested in shark blood chemistry. We happily spent all-nighters catching Spiny Dogfish from fishing barges off Redondo Beach for Dr. Urist. We were collecting sharks for science! We were on top of the world.

I entered college at UCLA, intent on a career that would lead to developing an effective shark repellant. That intent evaporated when I realized I was not a gifted chemist, or at least not a very good chemistry student. I completed my undergraduate studies, as a zoology not a chemistry major, at UC Berkeley, where I got more serious about sharks and fishes in general. I learned to scuba dive and managed the zoology department's fish collection. I took ichthyology (fish biology) from the late Dr. George Barlow, an animal behaviorist, which got me started observing fish behavior. Upon graduating, I joined the Peace Corps in Palau, Western Caroline Islands, as a fisheries specialist. Palau has spectacular coral reefs, and I spent most work hours and every weekend diving, catching, and watching fishes. I had my first up-close-and-personal encounters with a variety of reef sharks, often in the form of keeping them from stealing fish I had speared. I also was fortunate enough to serve as a local guide and field assistant to Dr. John (Jack) Randall of the Bishop Museum in Hawaii, a world expert on coral reef fishes. Jack and I published two papers on fishes in Palau, one on lo-

cal fish names and a second on the biology of and attacks by Blacktip Reef Sharks in the journal *Pacific Science*. I was now a bona fide shark researcher!

What followed was a master's degree from the University of Hawaii on land crabs and a Ph.D. from Cornell University on the behavior and ecology of lake fishes. Sharks had to be put on hold as other opportunities and interests developed. I was then hired at the University of Georgia in Athens, Georgia, as an ichthyologist, where I taught ichthyology, animal behavior, and conservation biology for 30 years. Although I studied marine and freshwater fishes, mostly by diving and observing predator-prey interactions, I always managed to get in as much shark biology and conservation as I could into my classes.

In essence, this book is really a boyhood dream come true. I can honestly say that I've fantasized about writing a book about sharks for as long as I can remember. In the meantime, I've published more than 50 research papers on various fishy topics, numerous book chapters, an ichthyology textbook (*The Diversity of Fishes: Biology, Evolution, and Ecology*, published by Wiley-Blackwell, with three co-authors), and a fish conservation book (*Fish Conservation: A Guide to Understanding and Restoring Global Aquatic Biodiversity and Fishery Resources*, published by Island Press), as well as co-authored with Bruce Collette of the Smithsonian Institution another book (*Fishes: The Animal Answer Guide*) in the same Johns Hopkins University Press series as this book. But really, all along, I wanted to write a book about sharks and was delighted when the Press invited me to do so, with George Burgess as my co-author.

From George Burgess
Like Gene, I have had a lifelong interest in sharks. The son of a career Air Force parent, I moved around a lot as a kid, as my father and our family were transferred about every three to four years. Luckily for me, those assignments were always at coastal locations, sequentially landing us in Hawaii, Virginia, Italy, and New Hampshire before we settled in on Long Island, New York. My father's love of the sea, coming from the angle of boats, seamanship, and swimming, opened the door for me to get hooked on its biota. We always had a family boat, and weekends and summer vacation activities frequently involved time at sea or on the beach.

I suffered mightily from asthma in my youth, at a time when medical treatment for the condition was all but lacking (and my parents smoked until I was 10 years old, a killer for asthmatic kids). Because breathing was a difficult task, I was especially reluctant to stick my head in the water. But while we were living in Naples, Italy, in 1957, skin diving, usually involving spearfishing, had become a developing pastime for younger guys. Noting my great interest in the writhing and dead critters being delivered to shore

by local divers and in an effort to teach me to swim, my father bought me a mask and snorkel. The seas parted, and I could see what lived down there! Breathing issues moved to the back burner, and my parents' problem became, not getting me in the water, but getting me out. When I wasn't in the water, I was walking the beaches or probing tide pools, looking for prizes.

As I got older, fishing became more important. My father manned the wheel, and I deployed the fishing gear. I'd set the hook, and we'd alternate reeling in the catches. Often, he'd bring some of his buddies with him, and they'd share their fishing knowledge with me. He noticed I wasn't as interested in the benchmarks of the other anglers—numbers and sizes of the catch—but was concentrating more on the diversity of species captured and the morphology of those fishes. An ichthyologist was incubating. My first shark catch was the mighty Spiny Dogfish, two feet of terror.

My first biology class in ninth grade sealed the deal. Blessed in having a dynamic teacher, Ernie Ernest, who was more fish than human, I discovered that I could turn my passion into a career, as had he. From there on, my goal was to become a biologist. He told me stories of close encounters he had with Sand Tigers while diving in nearby Long Island Sound and about their aggressive nature around speared fishes, my initial insight into shark attack. I needed to graduate into scuba, which I did my freshman year at the University of Rhode Island. Subsequent educational stops took me to the University of North Carolina and the University of Florida, where I studied oceanography and zoology. While in North Carolina I was able to finally work with "real sharks"—the kind seen so commonly today on TV and in the print media: Tigers and Whites, makos and Sand Tigers, Bulls and Blacktips. Research cruises in the Atlantic and the Gulf of Mexico introduced me to deepwater sharks like lanternsharks, miniature bioluminescent sharks that made my first Spiny Dogfish look like a giant.

After taking a position at the Florida Museum of Natural History at the University of Florida, I was fortunate that my scientific hero, Stewart Springer, retired to Gainesville. Despite a 40-year difference in ages, we became fast friends, most of my work days ending with a visit to his house for a couple hours of shark talk and collaboration on studies of lanternfishes and their kin. Stew had more hands-on experience with sharks than anyone of his or my generation, and I tried to absorb as much of it as I could. There never could be a better mentor.

My interests in sharks and rays are diverse. Colleagues, students, and I have collaborated to study their reproduction, food habits, age and growth, ecology, and movements; and we have discovered new species, including the two smallest species of sharks. We are deeply involved in fishery management and conservation, and as curators of the International Shark Attack File, we investigate and study shark attacks. The plight of sawfishes is

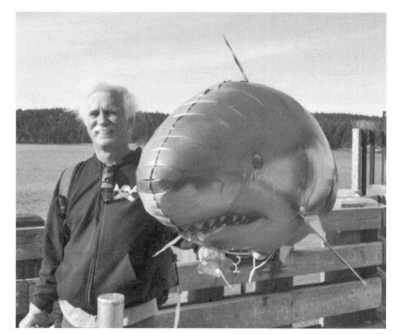

Gene Helfman waits at the Shaw Island, Washington, ferry dock with Cary, his helium-filled, radio-controlled White Shark. Gene and Cary lecture regularly to school and community groups about sharks and shark conservation. Cary is very adept at keeping people's attention.
Photo by Stephanie Buffum

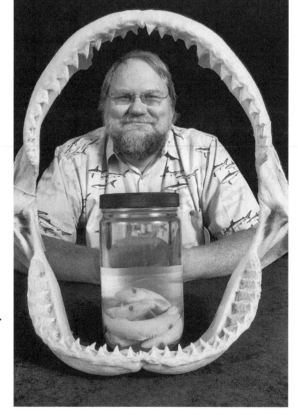

George Burgess with Dusky Shark jaws and preserved shark "pups" in his lab at the Florida Museum of Natural History. Large duskies are now rarely caught in the Gulf of Mexico, and museum specimens like these unfortunately document the way the natural world used to be. With proper fishery management, duskies and other sharks can return to their former level of abundance within our lifetimes. Photo by Florida Museum of Natural History

of particular interest, and we maintain the International Sawfish Encounter Database, which documents the occurrence of these endangered species. I am proud to have been a founding member of the American Elasmobranch Society and the IUCN Shark Specialist Group and have served as a president and vice-chair of those organizations. Our interest in shark attacks has naturally generated a lot of media interest, which opens the door to sharing thoughts on more meaningful shark-related issues. Gene and I hope this volume will serve to debunk some myths and spark additional interest in these fascinating animals.

And from Both of Us
Together, we have reflected on how great it would have been to have found a book such as this one when our fascination with sharks was new. Springer and Gold's excellent 1989 volume *Sharks in Question* came out about 25 years too late to educate us in our formative years. We hope this present book answers questions and sparks the kind of excitement about these wonderful animals that we have felt for so long, in readers from youngster to adult.

Sharks: *The Animal Answer Guide*

Chapter 1

Introducing
Sharks, Skates, Rays,
and Chimaeras

What is this book about?

This book is about chondrichthyan fishes (*chondros* + *ichthys*, literally "cartilaginous fishes"), a term that includes the subgroups of elasmobranchs (sharks, skates, and rays) and holocephalans (chimaeras), and is organized around common and not-so-common questions about these spectacular animals.

As a shorthand device and to avoid continually saying "sharks, skates, rays, and chimaeras," we refer to the overall group of chondrichthyans as "sharklike fishes," at least where all subgroups share a trait. Where differences occur or unique traits deserve mention, we are more specific. To avoid confusion, we use the technical term "selachians" (or "true sharks") when we mean sharks in the sense that most people think of them (the torpedo-shaped predators with prominent fins and big teeth that appear on the Discovery Channel). The term "batoids" refers to the more flattened skates, rays, and sawfishes (think of bat rays if that helps), and "chimaeras" is the general term for the strange-looking holocephalans. Also, ichthyologists (fish researchers) have agreed that common names of species are capitalized, so we talk about "Whale Sharks" and "White Sharks," but we don't capitalize "reef sharks" and "dogfish sharks" because those names refer to groups, not just one species.

The questions posed in this book often seem simple and elementary (for example, "Do sharks have to keep swimming to breathe?"). The answers, however, are often anything but simple because of the diversity and complexity of body types, species, ecology, anatomy, behavior, and physiology and the special solutions that sharks (and skates and rays and chimae-

1

Glossary of terms used with chondrichthyan (sharklike) fishes

Batoids	Skates, rays, guitarfishes, and sawfishes
Cartilage	Non-bony but calcified skeletal material
Claspers	Paired cartilage-supported male reproductive organs derived from pelvic fins
Chimaeras	Odd relatives (Holocephali) of the elasmobranchs; also known as ratfishes, rabbitfishes, and ghost sharks
Chondrichthyans	All sharklike fishes, including selachian sharks, batoid skates and rays, and chimaeras
Denticles	The scale type in sharklike fishes; technically, they are enamel-covered structures and are referred to as placoid scales
Elasmobranchs	Sharks, skates, and rays (but not chimaeras) with a slightly calcified cartilaginous skeleton, internal fertilization via modified pelvic fins, rows of replaceable teeth embedded in the connective tissue of jaws rather than in the jaw "bones" themselves, and scales (denticles) modified from tooth forerunners
Gill slits	External gill openings, typically five but as many as seven in elasmobranchs, one in chimaeras
Guitarfishes	A family of elongate rays with pronounced, pointed snouts
Rajiforms	The order of batoids made up of guitarfishes and skates
Rays	Several families of batoids that bear live young and have a long, whiplike tail, often with poisonous barbs
Sawfishes	A family of large, elongate rays with very pronounced, tooth-studded snouts
Sawsharks	A family of small, elongate sharks with very pronounced, tooth-studded snouts
Selachians	The group of elasmobranchs that includes sharks and sawsharks but not the batoid skates and rays
Skates	A family of rajiforms that lay eggs in cases ("mermaid purses") and have a thick tail with two small dorsal fins and a tail fin

ras) have evolved during their 400-million-year history. It is also worth remembering that these answers are the result of careful research conducted by shark biologists over the decades.

The chondrichthyan fishes include several groups whose names can sometimes be confusing at first. The table provides a quick listing of terms applied to the different groups of these related but diverse fishes and the traits they share.

Sharks: The Animal Answer Guide

What are sharks and how are they classified?

What then is a shark? The taxonomic class Chondrichthyes includes sharks, skates, rays, and chimaeras because they share a number of common traits. They are all fishes (aquatic animals with a vertebral column that breathe using gills and have scales and fins). All Chondrichthyes have a cartilaginous skeleton, which is a skeletal support system softer than bone because it is not as hardened with calcium as is true bone, although the skeleton is moderately hardened with calcium crystals. All groups also breathe via gills by pushing or pulling water in through their mouths and out their gill slits (with some minor exceptions). Because of anatomical differences, sharks, skates, and rays are grouped in one subclass, the Elasmobranchii ("plate gills"), and chimaeras are grouped in a different subclass, the Holocephali ("whole heads"). (The taxonomic names are shortened when used as adjectives or nouns—e.g., subclass Elasmobranchii, the elasmobranch sharks.)

Taxonomists and systematists, scientists who study and classify animals into groups on the basis of their characteristics and evolutionary relationships, have organized sharks into a hierarchy of groups that reflects those relationships. With over a thousand animals plus fossil forms to consider, not all systematists agree 100% of the time on the best classification, and new discoveries require modifications of older ideas.

In this book we generally follow the classification used by Joseph Nelson in the fourth (2006) edition of his internationally accepted *Fishes of the World*. Nelson has put together a classification from various authorities, chiefly from the work of Leonard J. V. Compagno. (The background publications are listed in the bibliography of Nelson's book for anyone who wants to explore the details.)

In this classification, elasmobranchs are divided into standard taxonomic groups and subgroups, reflecting their evolutionary relationships. The major groups, going from larger, more inclusive groups (infraclass, division, subdivision) to smaller, more exclusive taxonomic groups (superorder, order, family) are as follows, listed in phylogenetic order, which means that more primitive groups are listed first (additional details are given in appendix A):

Class Chondrichthyes (cartilaginous fishes)
 Subclass Elasmobranchii (sharklike fishes)
 Infraclass Euselachii (sharks and rays)
 Division Neoselachii (modern sharks, to distinguish them from
 the many, very different, extinct taxa)

Subdivision Selachii (sharks)

Superorder Squalomorphi

Order Hexanchiformes—two families of frill and cow sharks

Order Squaliformes—six families of dogfish, gulper, lantern, sleeper, and kitefin sharks

Order Pristiophoriformes—one family, saw sharks

Order Squatiniformes—one family, angel sharks

Order Echinorhiniformes—one family, bramble sharks

Superorder Galeomorphi

Order Heterodontiformes—one family, bullhead and horn sharks

Order Orectolobiformes—seven families of carpet, bamboo, Zebra, nurse, and Whale sharks

Order Lamniformes—seven families of Sand Tiger, Goblin, Megamouth, thresher, Basking, and mackerel sharks

Order Carcharhiniformes—eight families of cat, hound, requiem, and hammerhead sharks

Subdivision Batoidea (skates and rays)

Order Torpediniformes—two families of torpedo electric rays

Order Pristiformes—one family, sawfishes

Order Rajiformes—four families of guitarfishes and skates

Order Myliobatiformes—eleven families of stingrays, butterfly rays, eagle and manta rays

Subclass Holocephali

Superorder Holocephalimorpha

Order Chimaeriformes—three families of chimaeras, ratfishes, and ghostfishes

Differences among and between the various orders involve descriptions of their anatomy, including number and placement of fins, presence of fin spines, tooth number and characteristics, presence of a spiracle (modified gill opening on the top of the head for taking water into the gills), type of large intestine, number and position of gill slits, and presence of an eyelid—among other characteristics. A group-by-group breakdown requires too much detail for the present discussion, although basic body parts used in describing and distinguishing among species are shown in the illustration. One important anatomical feature that separates orders, families, and

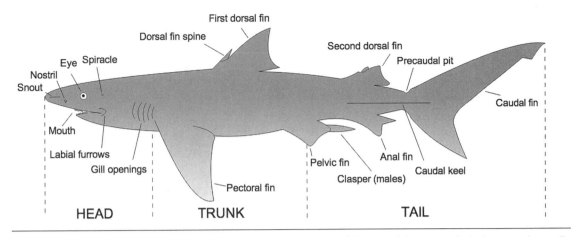

Body parts of a generalized shark. Many sharks lack spines in front of the dorsal fins; some lack the second, usually smaller dorsal fin; and a few have six or seven gill slits. Wikimedia Commons, http://en.wikipedia.org/wiki/File:Parts_of_a_shark.svg

even species is the shape and nature of the cartilaginous skull (chondro-cranium). Anyone interested in this important subject should consult the general references listed in the bibliography.

But several differences and characteristics are worth exploring. For example, all elasmobranchs have multiple external gill slits for expelling the water they breathe, whereas the holocephalan chimaeras have a single external gill slit. The gill arches are the internal cartilage structures on which the gills are supported. Water flows in the mouth, through the gill arches, and out the gill slits. The vast majority of sharks have five gill arches and five gill slits. But variety is the spice of life, and a few sharks and rays depart from the norm. Fortunately, this departure is usually reflected in their names, both common and scientific. Hence, sixgill sharks are *Hexanchus* (Greek for six openings), sevengill sharks are *Heptranchias* (seven openings), and the sixgill stingrays are in the family Hexatrygonidae ("six" plus a root word for stingrays that refers to their harplike shape; the adjective and noun form would be "hexatrygonid"). The Sixgill Sawshark, *Pliotrema warreni*, is not as conveniently named.

All chondrichthyan sharklike fishes have internal fertilization, the males possessing paired penis-like structures called claspers made from modified pelvic fins. In chimaeras the claspers have four lobes. Rays and most sharks are live-bearers, whereas some sharks and all skates and chimaeras are egg-layers. All sharklike fishes produce young that are miniature versions of the adult, which is to say that the young don't have a larval life stage. All holo-cephalans are marine, but many sharks, skates, and rays spend part or all of their lives in brackish or fresh water.

Another difference between the two main groups—elasmobranchs and holocephalans—involves their teeth, or dentition. Elasmobranchs have many teeth and replace old teeth with new teeth throughout their lives.

Introducing Sharks, Skates, Rays, and Chimaeras 5

A close-up photo of shark skin, showing the dermal denticles, or placoid scales. The head of the shark would be toward the top of the photo. Streamlining is increased by the ridges on the scales and their flared nature, each scale consisting of a flattened top sitting on a pedestal. NOAA (National Oceanic and Atmospheric Administration) Teacher at Sea Program, http://teacheratsea.files.wordpress.com/2011/08/dermal-denticles-magnified.jpg

The chimaeras have beaklike tooth plates that are worn and regrown as the animal grows. Their buck teeth and large eyes give many chimaeras a rabbitlike appearance, which is why one alternative name for the group is rabbitfish (not to be confused with the Siganidae, a family of tropical reef fishes also called rabbitfishes). Holocephalans are also called ratfishes and ghost sharks.

Elasmobranchs are more or less covered by scales called denticles that look like elongate plates with microscopic ridges, curved spikes, or pedestals protruding up from the surface. These denticles (technically, "placoid scales") may not overlap like the round, flat scales of bony fishes such as bass or perch or goldfish but are instead scattered around the body and packed together. Studies of developing embryonic sharks show that the scales are derived from the same tissues as the teeth in the mouth of a shark, whereas scales in bony fishes develop from bony layers in the skin. Holocephalans and many batoids are almost scaleless. All the chondrichthyan fishes except manta rays have their mouths and nasal (nostril) openings on the underside of their head; in bony fishes, the mouth is typically at the front end of the head and the nostrils are on the upper snout.

The major groups also differ in the way they swim. Sharks (but not skates and rays) swim by passing waves of muscular contraction from their head down to their tail, or by swinging their tail back and forth (see "How do sharks swim?" in chapter 2). Skates and rays swim using their pectoral (paired side) fins. Skates pass waves of contraction along the outer edges of these fins; rays similarly pass waves along the edges but also flap their large pectoral fins, much like birds flap their wings. Chimaeras use both

A juvenile Pacific Spookfish, *Rhinochimaera pacifica*. This relatively large chimaera can grow to 1.6 m (5.3 ft) long. The fleshy snout, whose function remains a mystery, becomes more narrow and pointed in adults. Photographer unknown

their bodies and their fins to swim, passing a wave of muscular contractions down the body but also flapping their pectoral fins. When not moving, fishes need a way of keeping from sinking in the water. Bony fishes have a gas-filled bladder for maintaining buoyancy; chondrichthyans use a large, oil-rich liver to achieve the same end.

Chimaeras are chondrichthyan fishes but not elasmobranchs. They are in a separate subclass, the Holocephali, with one modern order, the Chimaeriformes (and six extinct orders known from fossils that are very dissimilar to modern forms). Aside from their cartilage skeletons, chimaeras share few similarities with sharks and rays, and they almost seem like a hodgepodge assembled from spare parts. The name, in fact, comes from Greek mythology, in which a chimaera was a monster constructed of unlikely parts.

Sharks can be told apart from skates and rays fairly easily. A major difference is that in selachian sharks, the gill slits are on the sides above and ahead of the pectoral fins, whereas in batoid skates and rays, the gills are on the belly (ventral) side below the pectoral fins and behind where the front end of the pectorals attach to the body. Also, most sharks have a relatively narrow attachment point for their relatively narrow pectoral fins, and the front edge of the pectoral fin is attached behind the head. In batoids, the pectorals are broadly attached, sometimes as wide as the body is long; and the front edge of the pectoral fin may be attached to the head, giving the fins a winglike appearance and producing an overall disklike shape to the body. Chimaeras' pectoral fins are not as broad as those of rays, and the attachment to the body is just behind the head and quite narrow, more like that of a bony fish such as a seabass. Teeth also differ among groups. Batoids have flat, pavement-like teeth that carpet the mouth, which means multiple rows function to crush prey. Selachians usually have more pointed teeth around their jaws, with the outer row the most functional. Chimaeras' teeth are fused into a beak. Although body shape would seem an obvious identifier—selachians are torpedo-shaped and round in cross-section,

and skates or rays are flat from top to bottom—too many exceptions occur to make that a reliable distinction.

Telling selachian sharks apart from batoid skates and rays is relatively easy (with some exceptions noted below), but telling skates (rajiforms) apart from rays (myliobatiforms) can be more challenging. In general, skates lay eggs that have a horny shell (often called a "mermaid's purse"), whereas rays give birth to live young. Both skates and rays have long, skinny tails that are separate from the body (in selachians the tail is thick and continuous with the body), but in skates the tail is thicker and has two small dorsal fins and a caudal (tail) fin at its end. In rays the tail is long and whiplike and has barbs (stings) on the top near the base. Stingray barbs develop embryonically from scales, so ultimately they are another kind of teeth.

And then we have elasmobranchs that appear intermediate between groups: the selachian wobbegongs, angel sharks, and saw sharks; and the batoid sawfishes and guitarfishes. Wobbegongs (Orectolobidae), also called carpet sharks because of their shape, color, and bottom-living habits, are true sharks (selachians). They are highly flattened top-to-bottom (dorsoventrally depressed). But the gills are on the sides, the pectorals are attached behind the head, and the teeth are fanglike, all good "shark" characteristics. Saw sharks (Pristiophoriformes) and especially angel sharks (Squatiniformes) are also dorsoventrally flattened. Angel sharks are even referred to as "raylike" in technical books, and their scientific name actually comes from the Latin word for skate. But again, an angel shark is a true shark because its large pectoral fins are not fused to its head, the tail is thick and continuous with the body, the teeth are long and pointy, and it bears live young, unlike any skate. Confusing the issue a little is the location of the gills, which are below the pectorals, but we will have to ignore that deviation.

Saw sharks (Pristiophoriformes), as their name implies, are also true selachians. However, they have the look of batoid sawfishes (Pristiformes), not of rays or skates. In biological terms, they have converged during evolution on the body plan of a sawfish, having a long snout armed with teeth on the outside, which they apparently use to slash prey. In saw sharks, gill openings are on top (the shark condition), whereas in sawfishes they are under the pectoral fins, a batoid trait. Saw sharks never get over 150 cm (5 ft) long and are usually much smaller. Sawfishes are huge, up to 7 m (23 ft) long. Saw sharks live in relatively deep water, some at temperate latitudes, whereas sawfishes are shallow dwellers in tropical and subtropical areas and even live in fresh water.

The last, slightly confusing group contains the rajform guitarfishes, an intermediate between sharks and rays that are even referred to in some taxonomic publications as "sharkrays." Sharklike features include the tail

Saw sharks are true selachian sharks that have evolved the same basic body form as the sawfishes, which are actually batoid rays. Because saw sharks are relatively small deep-water animals, good photos of them are comparatively rare. (*Top*) Illustration of an Australian Shortnose Sawshark, *Pristiophorus nudipinnis*. (*Bottom*) Photograph of the head and saw of a South African Sixgill Sawshark in the British Museum of Natural History. Top, illustration from the *Illustrated Catalogue of the Fishes of South Australia* (Adelaide: G. Hassell & Son, 1921), courtesy of the Freshwater and Marine Image Bank, en.wikipedia.org/wiki/File:Pristiophorus_nudi pinnis.jpg; *bottom*, photo by Gene Helfman

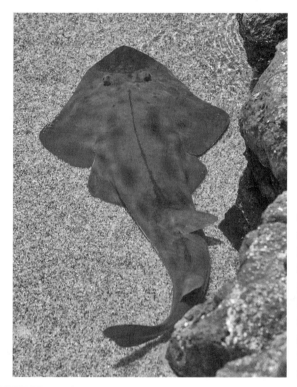

Guitarfishes are sharklike rays that spend their lives lying on the bottom, often in shallow water. This Shovelnose Guitarfish (*Rhinobatos productus*), a common species along the California coast, shows well where the common name for the group comes from. Photo by Jot Powers, Wikimedia Commons, http://en.wikipedia.org/wiki/File:Shovelnose_guitarfish.JPG

section, which is thick and continuous with the head and trunk, and the tail fin, which is large, as are the dorsal fins. But as with (other) skates and rays, the body is flattened, the pectoral fins are expanded and attached from the head back to midbody, and the gill openings are underneath.

How many kinds of sharks are there?

The most recent analysis of chondrichthyan diversity puts the number of sharks, skates, rays, and chimaeras at almost 1,300 species. Breaking this total down into its taxonomic parts, selachian sharks make up roughly 40% of the total, batoid skates and rays make up 58%, and chimaeras around 2%. Ground or reef sharks (of the order Carcharhiniformes, containing Blacktip Reef and Oceanic Whitetip sharks, Blue Shark, Tiger Shark, Lemon Shark, and others) constitute more than half the shark species. They are followed by the squaliform dogfishes (about 100 species); 32 orectolobiform carpet, nurse, and bamboo sharks; 15 lamniform mackerel sharks (makos, White, threshers, and Basking Shark); and 15 squatiniform angel sharks. About 20 species are in smaller orders (heterodontiform horn sharks, hexanchiform cow and frill sharks, echinorhiniform bramble sharks, and pristiophoriform saw sharks). (Recall that the taxonomic name for orders is shortened as an adjective or noun—e.g., order Lamniformes, the lamniform sharks.)

Among the batoids are about 300 species of skates and guitarfishes (order Rajiformes); 200 species of myliobatiform stingrays, butterfly rays, and manta or eagle rays; 60 torpediniform electric rays; and fewer than 10 pristiform sawfishes. Chimaeras (subclass Holocephali) include 46 species, all in the order Chimaeriformes (see appendix A for details).

The total of 1,300 species is not a final number because researchers find and describe a few new species almost every year. The number of discoveries surprises people unfamiliar with shark diversity: just in the first decade of the twenty-first century, more than 180 new species of chondrichthyan fishes were described. Most of these are deepwater elasmobranchs or animals from tropical regions that have not been as well studied as temperate waters. Twenty previously unknown sharks and rays were discovered during surveys of local fish markets in Indonesia, giving new meaning to the idea of adventure in shopping. Some species are discovered when we look in places we hadn't sampled before or when we "sample" by different methods. For example, a new species of swell shark, Steven's Swellshark (*Cephaloscyllium stevensi*), was described in 2011 on the basis of five individuals caught in a trap set for chambered nautilus (a mollusk) off Papua, New Guinea. The traps were set 250 to 275 m (820–900 ft) deep, well below the depths that can be reached by divers using even advanced, mixed-gas scuba

Sharks: The Animal Answer Guide

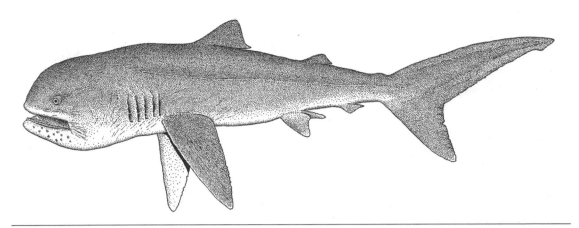

The 5-m-long (16-ft) Megamouth Shark was not discovered until 1976, and fewer than 60 specimens have been found, most washed up dead or dying on beaches. Who knows what other large shark species remain to be discovered? Drawing by P. Vecsei; used with permission

gear. The sharks were relatively small, about 0.75 m (2.5 ft) long. Swell sharks are unique among sharks because of their ability to gulp water into their stomach and swell up, probably an antipredator defense that is better known in bony pufferfishes.

Some discoveries involve spectacular, often larger animals, including a very large shark and the largest species of ray. In 1976, a very large new species of shark became entangled in the sea anchor of a naval ship near Hawaii. Now known as the Megamouth Shark, *Megachasma pelagios*, this 4- to 5-m (13- to 16.5-ft) open-ocean shark feeds on zooplankton and small fishes. Dead or dying individuals occasionally wash up on beaches in tropical and subtropical regions, but fewer than 60 specimens have been found. Megamouth is certainly the most spectacular shark species to be discovered in the past century.

In the case of large rays, scientists didn't find new animals but used behavior, ecology, and anatomy to reveal that what we thought was one worldwide species, the Manta Ray, *Manta birostris*, is actually two species, *M. birostris* and *M. alfredi*. Manta rays evolved from more "normal" stingrays and in the process lost the barbed sting characteristic of the ancestral group. The larger of these two mantas, Giant or Pelagic Manta, *M. birostris*, still has a tiny, nonfunctional ("vestigial") spine, whereas the Reef Manta, *M. alfredi*, has lost all vestiges of the sting. The two species also differ in coloration, migratory behavior, and affinity for shallow water and reefs.

How did sharks get their name?

Controversy surrounds a basic fact about sharks, namely, the origins of their common name. One possibility is a Mayan word, *Xoc*, probably

pronounced *shock*, which was the name of a demonic, fishlike god. Apparently, Europeans, prior to the exploratory ocean voyages of the sixteenth century, did most of their fishing in rivers and lakes or very close to shore. They had little experience with large oceanic sharks. When they eventually traveled to the New World, they were astonished by, and had no name for, the large, fiercely predatory sharks of the tropics. So they adopted the local word for these animals and carried it back to Europe. Another possibility is a German term, *Schurke*, which refers to a shifty criminal. However, the modern German term for "shark" is *Hai* or *Haifisch*.

Nusse is another old English name for sharks in general and may be the origin of the current names of various sharks referred to as nurse sharks (Nurse Shark, *Ginglymostoma cirratum*, and Grey Nurse, *Carcharias taurus*, also called Sand Tiger or Ragged Tooth). However, another possible origin for the Nurse Shark's name relates to its well-known ability to create suction pressure when feeding, as shown in laboratory studies. Nurse Sharks are thought to suck lobsters and other prey out of holes, although verification of this ability in nature is lacking. People other than researchers are familiar with a Nurse Shark's sucking habits. Local names for relatives of the Nurse Shark suggest this behavior. In the island nation of Palau, a tropical Pacific species, the Zebra Shark (*Stegostoma fasciatum*) is referred to as *metmut*. This is the same word used to describe the suckling behavior of nursing human infants. Sand Tigers, with their decidedly snaggletooth appearance, are equipped to impale prey fish on long, pointed teeth, so it is unlikely that their name has anything to do with sucking.

Where do sharks live?

Most sharks, skates, and rays and all chimaeras live in the ocean, from the poles to the tropics and from the shoreline out to the middle of the ocean. Among the elasmobranchs, depths range from exceedingly shallow water only a few feet deep, down to great ocean depths. Stingrays, guitarfishes, and sawfishes feed on the bottom near the shoreline, one of the reasons you should shuffle your feet when wading in shallow water where stingrays may occur. Lemon Sharks (*Negaprion brevirostris*), including large individuals 3 m (about 9 or 10 ft) long, will swim in water so shallow their dorsal fins and tail stick out of the water. Large groups of Blacktip Reef Sharks (*Carcharhinus melanopterus*) also swim in water shallower than their body depth. Chimaeras usually occur in water deeper than 80 m (250 ft), with some species found as deep as 2,600 m (8,500 ft).

Sharks in general do not live in the greatest depths of the ocean, especially when compared with bony fishes. Bony fishes have been caught in bottom nets in the Puerto Rico Trench at a depth of 8,370 m (27,500 ft).

Few sharks live below 3,000 m (9,840 ft), the record holders being a Portuguese Dogfish (*Centroscymnus coelolepis*) seen at 3,690 m (12,100 ft) and an unidentified dogfish (Squalidae) at 4,050 m (13,280 ft).

In terms of latitude and water temperature, selachian true sharks are most diverse in warm temperate oceans, although many occupy cold, deep water below the warmer surface. The tropics have the next highest diversity, followed by cold temperate and polar seas (see "Which geographic regions have the most species of sharks?" in chapter 5). Stingrays are more diverse in inshore tropical regions. In contrast, more species of skates are found in temperate oceans and in cool, deep waters than in the shallow, warm tropics. Chimaeras are also chiefly cool-water species.

Some selachians swim near the surface, others prefer the bottom, and other species are found at midwater depths. Skates and rays are almost entirely bottom-livers, feeding on organisms that live on or in the bottom sediments. The notable exceptions are the manta rays, which swim in midwater and feed on zooplankton (small invertebrates up in the water column). Chimaeras live near the bottom and also feed on bottom-dwelling invertebrates that they crush with their tooth plates.

A few rays and even fewer sharks live and reproduce in fresh water, including far up major rivers (see "How many shark species live in rivers and lakes?" in chapter 5).

When did sharks evolve?

Fossil teeth that undoubtedly belonged to sharks have been found in geological deposits that date to the early Devonian period, about 418 million years ago. An actual sharklike fossil with the revealing name of *Doliodus problematicus* is thought to be 409 million years old. (Fossils of entire sharks are very rare because the cartilage that makes up most of a shark's skeleton does not fossilize well; we usually have just teeth to tell us about the entire animal.) Some sharklike scales from the late Ordovician (455 million years old) have been found, but scientists are hesitant to proclaim that they came from a shark. Holocephalans arose somewhat later, the oldest fossils dating back to at least the late Devonian, about 375 million years ago. The batoid skates and rays appear even later, in the Jurassic (about 175 million years ago). The earliest chondrichthyans already had internal fertilization (evident as claspers on apparent male specimens) and replacement teeth.

Sharks were much more diverse in body shape and habitat in the past. In the late Paleozoic and Mesozoic periods (300 to 100 million years ago), sharks were the dominant predators in both marine and freshwater habitats, instead of being primarily marine as they are today. Some early sharks were *Cladoselache, Xenacanthus, Ctenacanthus, Hybodus,* and *Helicoprion. Helicoprion*

Helicoprion bessonovi, an edestoid shark that lived in the late Paleozoic and early Mesozoic eras (300 to 200 million years ago). It is thought to have grown to 3 to 4.5 m (10 to 15 ft) in length. Its "circular saw" jaw and tooth arrangement are unlike those of any living shark, and its body shape remained a mystery for many years. **(A)** Fossilized remains of the lower jaw, about 26 cm (10 in) in diameter; **(B)** hypothetical reconstruction of the animal. *A*, from J. Walther (1914), Wikimedia Commons, http://en.wikipedia.org/wiki/File:Helicoprion_Bessonovi2.jpg; *B*, drawing by Dmitry Bogdanov, Wikimedia Commons, http://en.wikipedia.org/wiki/File:Helicoprion_bessonovi1DB.jpg

("spiral saw") had a lower jaw and tooth form unlike any living shark, its teeth arranged in a spiral of replacement teeth rather than a toothy grin. Until a more complete fossil of the head of a relative was found, researchers did not know what *Helicoprion* was (some thought it was a relative of the chambered nautilus, a mollusk) and where on the body the tooth spiral was located.

What are the largest and smallest sharks alive today?

Before an answer to the question of large sharks can be given, we have to admit that very large sharks are hard to catch and notoriously difficult to measure accurately. Many accounts estimate (or guesstimate) size, with characterizations such as "bigger than the boat," or try to deduce length from photographs. Weights are even less often accurately measured because scales capable of dealing with animals that weigh many tons aren't always available when a super-large shark is hauled to a dock or boat ramp. As a result, verified maximum lengths are often shorter than suspected maximum lengths, and exaggeration is not unknown.

Recognizing these limitations on accuracy, we know which species are largest, but we don't really know how big they get. The largest shark—and for that matter, the largest fish alive today—is the Whale Shark, *Rhincodon*

typus (in its own family, the Rhincodontidae). Whale Sharks have been measured at 12.6 m (41 ft) long and 12,000 kg (26,500 lb) in weight. A Whale Shark caught off Anping, Taiwan, in 1994 reportedly weighed 35,800 kg (79,000 lb) and appears to be the largest specimen ever recorded. Larger individuals have been seen or caught, with reported lengths of 16 m (55 ft) and 18 m (60 ft), but these claims have not been verified. An 18-m (60-ft) Whale Shark could weigh 34,000 kg (75,000 lb), so the Anping shark could have been close to that length. Regardless, these sizes make Whale Sharks the longest shark that has ever lived. The second largest shark is the Basking Shark, *Cetorhinus maximus* (also in its own family, the Cetorhinidae), which may get to be 10 m (33 ft) long and 4,000 kg (8,800 lb). Both Whale and Basking sharks feed mainly on plankton and small fishes.

Sleeper sharks (*Somniosus*) are also very large sharks. Pacific Sleeper Sharks (*Somniosus pacificus*) are reported to reach lengths of up to 7.6 m (25 ft), but the largest one measured was 4.4 m (14 ft) long. The Greenland Shark, *Somniosus microcephalus*, is thought to grow to 6.4 m (21 ft) and 1,000 kg (2,200 lb), and possibly up to 7.3 m (24 ft). Tiger Sharks, *Galeocerdo cuvier*, come close, with the largest reliably measured at 5.5 m (18 ft) in length (estimated weight 1,418 kg, or 3,120 lb), although another Tiger Shark, caught in 1957, was reported to be 7.4 m (24 ft) long and weighed 3,110 kg (6,860 lb). Great Hammerhead Sharks, *Sphyrna mokarran*, also grow large, with the longest reported length at 6.1 m (20 ft) and a recognized world record weight of 450 kg (991 lb).

And of course there is the White Shark, *Carcharodon carcharias*, about which misinformation abounds. Many "true accounts" of giant White Sharks have appeared over the years, with published sizes (in newspapers) of up to 36.5 ft (11.1 m; that one was a typographical error—the fish was actually only 16.5 ft, or 5 m, long). Other claims for 9-m (29-ft) and 7-m (23-ft) White Sharks have also turned out to be exaggerations. The largest White Shark reliably measured was either a 5.9-m-long (19.5-ft) female caught off Ledge Point, Western Australia, in 1984 or a 6.1-m-long (20.3-ft) female captured off Prince Edward Island in August 1983. The heaviest reliably measured White Shark was a 3,427-lb (1,560-kg), 17-ft (5.2-m) female captured off Montauk Point, New York, in 1986. (A 20-ft- or 6.1-m-long shark would weigh closer to 4,200 lb, or 1,900 kg.) Based on the width of bite marks on dead whales, and recognizing that mouth width is a good (but not exact) indicator of body length, modern White Sharks may reach 6.4 m (21 ft) long, and some reasonable accounts put that number at 7 m (23 ft). However, none that large have been caught and carefully measured, despite what you may read in popular books, in magazines, and on websites.

At the other end of the size spectrum are some decidedly small sharks,

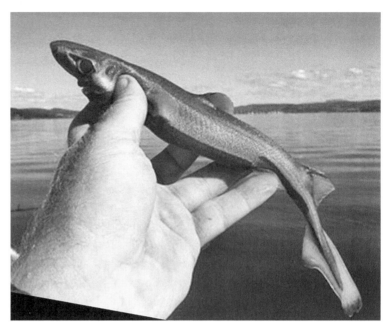

The world's smallest shark, the 17-cm, 15-g Dwarf Lantern Shark, from deep waters off northern South America. Photo by Javontae Murphy, Wikimedia Commons, en.wikipedia.org/wiki/ File:Dwarf_Lanternshark.jpg

the smallest being the Dwarf Lantern Shark, *Etmopterus perryi*, at 17 to 21 cm (7–8 in) and 15 g (0.5 oz), and several that are slightly larger, measuring 21 to 25 cm (9–11 in). These include the Cylindrical Lantern Shark, *Etmopterus carteri*; two dalatiids, the Spined Pygmy Shark (*Squaliolus laticaudus*) and the Smalleye Pygmy Shark (*S. aliae*); and the proscyliid Pygmy Ribbontail Catshark (*Eridacnis radcliffei*). All of these small sharks live in very deep water.

Among batoids, the largest sawfish is the Narrowsnout Sawfish (*Pristis zijsron*) of the Indo-West Pacific, reliably measured at 7.3 m (24 ft) long, with an estimated weight of 1,418 kg (3,120 lb). The Smalltooth Sawfish, *Pristis pectinata*, may reach 9.4 m (30.8 ft) in length and weigh 2,591 kg (5,700 lb). Two species hold the claim for largest stingray. The Giant Freshwater Stingray (*Himantura chaophraya*), of large rivers and estuaries of Southeast Asia, can be 4.6 m (15 ft) long, 1.9 m (6.2 ft) across, and 500 to 600 kg (1,100–1,300 lb). *H. chaophraya* also has the largest barb of any stingray, at 38 cm (15 in) long. The largest marine stingray is the Smalleye Stingray (*Dasyatis microps*) of Indo-Pacific reefs, which can reach a width of 2.2 m (7.2 ft) and a total length of 3.2 m (10 ft). The largest ray is the Giant Manta Ray, reputed to measure up to 6.7 m (22 ft) from "wingtip" to "wingtip" and weigh in excess of 1,360 kg (2,290 lb). Among the smallest rays are some gymnurid butterfly rays, at only 30 cm (12 in) long. The Finless Sleeper Ray (*Temera hardwickii*) of Southeast Asia is even smaller, maturing at 8.2 cm (3.2 in) long and seldom exceeding 15 cm (5.9 in) in length, which would make it the smallest chondrichthyan anywhere. The

Sharks: The Animal Answer Guide

smallest skates, the pygmy skates, are in the genus *Fenestraja*; they reach total lengths of 20 to 30 cm (8–12 in).

Chimaeras are more modest in size, with an adult length of 40 to 150 cm (1.3–4.9 ft); females are often larger than males, but not in all species. The largest chimaeras are Carpenter's Chimaera (*Chimaera lignaria*) from New Zealand and the Rabbit Fish (*Chimaera monstrosa*) from the eastern Atlantic. At the shorter end of the range is the 40-cm (16-in) Striped Rabbitfish, *Hydrolagus matallanasi*, from coastal Brazil.

These size comparisons highlight an interesting fact about variation among selachian families. Large (or small) size is not necessarily a family trait. The somniosidae sleeper sharks and dalatiidae kitefin sharks are two very closely related squaliform families, so close that some taxonomists consider all of them in one family, the Dalatiidae. This group includes sharks that fall into several extreme categories. Two dalatiids are among the smallest sharks known, at 22 to 25 cm (9–11 in) long, whereas the somniosid sleeper sharks are among the largest, reportedly 7.3 to 7.6 m (24–25 ft) long. One of these, the Greenland Shark, holds the record for farthest-northern-occurring shark, being captured along the northern coast of Greenland, above 83° north. (The North Pole is at 90° north, or only about 400 miles north of northern Greenland.) The Portuguese Dogfish is a somniosid and is the deepest-living shark known.

Finally, and to the disappointment of 14-year-old boys everywhere, the biggest sharks are usually female.

What is the fastest shark?

Selachian true sharks on the whole are slow swimmers. Their form of swimming, which involves passing S-waves down the body from head to tail, limits their speed compared to that of bony fishes (see "How do sharks swim?" in chapter 2). The fastest bony fishes are tunas and their relatives. Tunas swim by quick side-to-side beating of the tail. The fastest sharks do the same and are aptly named mackerel sharks because of this convergent swimming motion. The fastest shark is the Shortfin Mako, *Isurus oxyrinchus*. Makos have been clocked at speeds of 32 kph (20 mph). This may not sound impressive, which may be why you will find erroneous swimming speeds on websites ranging as high as 95 kph, or 59 mph. Nothing— manmade or otherwise—goes that fast underwater for very long. Nuclear submarines top out at 46 kph, or 29 mph. Remember: a shark only has to swim as fast as or slightly faster than its prey, for a short period. A study of swimming in Yellowfin Tuna (*Thunnus albacares*), one of the mako's favorite foods, showed average burst speeds of 24 mph, with a range of values for the four fish studied of between 11 and 46 mph. The mako doesn't have

to catch the fastest tuna, anymore than the cheetah has to catch the fastest gazelle. Slower ones are just as tasty.

What is the largest shark that ever lived?

If we disregard the speculated lengths of the living Whale Shark, the largest true shark in terms of weight to swim the seas (or lakes) occurred around much of the world during the mid-Miocene through the Late Pliocene, 16 to 1.6 million years ago. This creature was *Carcharocles megalodon*, the Megatooth Shark. Recent fossil evidence has confirmed the need to change the Megatooth's genus name from *Carcharodon* to *Carcharocles* because we now know that modern White Sharks—*Carcharodon*—evolved from mako shark ancestors, not Megatooth ancestors. Researchers previously thought a direct connection existed between the Megatooth and White Sharks because of tooth similarities, but new fossil findings confirm separate ancestries and therefore identities (although not all experts agree on this change).

Giant teeth from the Megatooth Shark are found in fossil deposits on land and in the sea in Europe, Africa, Australia, India, Japan, North and South America, and some western Pacific islands. The largest teeth ever found came from South Carolina and Peru and measured 17.6 to 17.9 cm (7.25–7.37 in) from the tip to the end of the root, or base, with an enamel height of 16.8 cm (6.9 in). Paleontologists use a formula for calculating shark length based on tooth size, but they use enamel height rather than total tooth size because teeth vary greatly in root shape. They also make the reasonable assumption that the proportions of the Megatooth Shark are similar to those of the White Shark. A Megatooth Shark with an enamel height of 168 mm would be about 16 m (52 ft) long and weigh approximately 48,000 kg (105,000 lb). The jaws of such a monstrous shark would have been more than 1 m (3.25 ft) wide, the dorsal fin would be 1.4 m (4 ft) high, and the tail would be 1.75 m (6 ft) tall. Several people have reconstructed the jaws using entire sets of fossilized Megatooth teeth.

The Megatooth Shark, although probably the largest predatory shark that ever lived, occurred with other relatively gigantic 5- to 6-m long (16–20 ft) predators during the same time, including the Speartooth Mako (*Isurus hastalis*) and a hemigaleid, *Hemipristis serra*, a relative of the modern Snaggletooth Shark, *Hemipristis elongatus*. Megatooth Sharks also swam the seas with their smaller cousin, the White Shark.

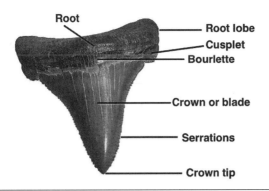

Root
Root lobe
Cusplet
Bourlette
Crown or blade
Serrations
Crown tip

Tooth of the extinct *Carcharocles angustidens* (Oligocene, 45–28 million years ago), showing basic terms applied to a shark's tooth. The crown is where enamel height is measured; cusplets are also called "side cusps." Redrawn from THobern, Wikimedia Commons, http://en.wikipedia.org/wiki/File:Lingual_side .JPG

Bashford Dean's famous 1909 reconstruction of the jaws of *Carcharocles megalodon,* displayed in the American Museum of Natural History. Dean was the first to attempt such a reconstruction, from which he estimated Meg's length at 30 m (98 ft). Unfortunately, the cartilage portions of both jaws were too tall, and this mistake affected the eventual length calculation. Recent reconstructions sell for around $10,000 if made with artificial teeth, $100,000 if real fossils are used (www.megalodonteeth. com). Maybe Santa has room in his sleigh. Wikimedia Commons, http:// en.wikipedia.org/wiki/File:Carcharodon_mega lodon.jpg

Where can I find fossil shark teeth?

Replacement teeth developed early during shark evolution, as is evident from the abundance of fossil teeth found in many parts of the world. Some of the richest deposits in the United States, in terms of numbers and species diversity, occur in coastal South Carolina; Calvert Cliffs State Park in Maryland; many locales in Florida, including Bone Valley in Polk County; Sharktooth Hill near Bakersfield and Scotts Valley near Santa Cruz, California; and several locales in North Carolina and New Jersey, among others. Teeth still embedded in sedimentary rock occur most commonly in

deposits 100 to 2.5 million years old (Late Cretaceous through the Tertiary period). Embedded teeth require careful digging and picking to free them from the surrounding rock. Individual, isolated teeth of all sizes come from animals that lived recently to about 25 million years ago and can be found in land deposits, by beachcombers just above the tide line, or by divers on the ocean floor. Two good sources of information and photos from many species and locales are at www.blackriverfossils.org and www.sharkteethrus .com. Fossil shark teeth are also a popular item on Internet auction sites.

Shark teeth aren't the only fossilized remains found in these areas. Stingray spines and tooth plates, vertebrae, sawfish "teeth," and shark feces also fossilize. Shark "coprolites" (fossil shark poop) are identifiable as having belonged to sharks because they have a twisted nature, having passed through the spiral valve intestine of the animal.

How can you tell whether a fossil shark lived in a river, a lake, or an ocean?

The kind of rock a fossil occurs in is a good indicator of what the animal's habitat was. Look in an ichthyology (fish biology) textbook and you will find statements such as, "In the Late Devonian, sharks were found primarily in freshwater habitats, in contrast to their primarily marine distribution today." Researchers frequently determine the habitats that fossilized aquatic animals lived in by studying the kinds of hardened sediments in which the fossils occur. Marine mud and sand have chemical and physical characteristics (particle size, organic matter, rock type) that are different from freshwater habitats, and lakes have different sediments from rivers. When sediments become hardened into rock over time, which is the process that creates fossils, the original differences result in different rock types.

Additional clues come from the other animals and plants found in the same deposits as the shark fossil. Some organisms only lived in the ocean (ammonites, trilobites, sea scorpions), others only in fresh water (horsetail rushes, amphibians). But sometimes the fossils get moved around and away from their original deposits, as happens in floods and mudslides. Fossils of freshwater animals then wind up in the ocean, confusing the picture. Determining the habitat type of a fossil is clearly a mystery-solving process, and the detectives who solve such mysteries (paleontologists, geologists), like any good detective, must have knowledge in a number of disciplines to complete the task.

Why are sharks important?

Sharks are important because of the roles they play in ecological systems. In all habitats, on land and in the water, the natural food web has multiple levels and connections, beginning with the plants that produce food from the sun, up to the predators that feed on plant-eating animals, the predators that feed on other predators, and the scavengers that feed on decomposing remains. Sharks occupy the top levels of food webs in all of the world's oceans. Such top-level carnivores keep the numbers of other species in check, preventing both starvation and overuse of resources. When top level carnivores are eliminated from a system, their absence leads to both overpopulation and depopulation at other levels because the impacts cascade down through the food web.

Predators also remove diseased, injured, or poorly adapted individuals from prey populations. This predatory behavior is a part of natural selection, constantly improving the genetic make-up of prey populations and even potentially reducing the likelihood that diseases will spread. Sharks also scavenge on the remains of dead animals, recycling the energy and nutrients that are locked up in decomposing bodies. Sharks in their roles as both predators and scavengers are essential to the health of marine ecosystems.

Why should people care about sharks?

Humanity depends on healthy ecological systems for food, clothing, medicines, building materials, clean water, and mental well-being. When fished at sustainable levels, sharks provide protein for people who might otherwise have a protein-deficient diet. Drugs made from sharks and based on the chemistry of sharks have been used to treat human bone diseases and cancer (see "Are any medicines made from sharks?" in chapter 8). When sharks are overfished and ecological systems are thrown out of balance, the direct and indirect usefulness of sharks to humans and other organisms is lost.

But it is not only the practical use of resources that affects our well-being. Many studies have shown that people get genuine pleasure from experiencing nature, especially wild animals in their natural surroundings. We have a deep-seated affinity for and curiosity about the natural world, in no small part because it has been evolutionarily advantageous to understand the living world around us. Our ancestors who learned about and understood nature were more likely to pass on their genes than early humans who were less informed. Ecotourism, the popular practice of traveling to

natural areas and learning about them in a nondestructive, nonconsuming manner, is a modern result of this affinity. A healthy environment is synonymous with a healthy human population. And sharks are among the first species to disappear when we degrade natural conditions in the ocean through overfishing, pollution, and habitat destruction.

Chapter 2

Form and Function of Sharks

What is the metabolism of a shark?

Sharks live in the slow lane. They have what we consider to be very slow metabolisms and are built to conserve energy in almost all aspects of their lives, especially when compared with us or even with bony fishes. Their fins and scales and body shape, as discussed in questions in this chapter, are optimal for moving through the water efficiently. In addition, their resting metabolic rates are comparatively low, about one-third that of similar-sized bony fish. Their active metabolic rates are only about 3 times that of resting rate, whereas in a bony fish, active rates are 10 times that of resting rates. Feeding rates are also low: a 2-kg (4.5-lb) Spiny Dogfish, *Squalus acanthias*, needs only 8 g (about a third of an ounce) of fish per day to maintain itself, whereas a similar-sized salmon needs four times that amount.

The above generalizations hold when we talk about small sharks such as the Spiny Dogfish, which are about the size of some bony fishes. It is considerably harder to measure the metabolism of large sharks because of difficulties of keeping them in captivity. But indirect measures can be used, such as estimating oxygen consumption by looking at body temperature as a shark with a transmitter passes through water of different temperatures. This was done for a 4.5-m-long (14.75-ft) White Shark. The resulting calculation indicated that the White Shark used about one-eighth of the oxygen per pound of body weight that a human being uses. The same calculations give us some idea of how often a White Shark must feed to maintain itself. If the same shark fed on a dead whale and ate 30 kg (66 lb) of whale blubber, which is not an unusual amount for a White Shark, it could survive for six weeks before it would have to feed again.

Are sharks cold-blooded or warm-blooded?

The great majority of sharks, rays, and chimaeras are cold-blooded, meaning their body temperature is the same as the surrounding water. However, despite the classic description of sharks as "cold-blooded killers," some sharks—including the highly predatory lamnid sharks (White, makos, Porbeagle, Salmon) and the closely-related thresher sharks (Alopiidae)—are actually warm-blooded: they maintain a body temperature that is warmer than the water around them.

Muscle activity warms the blood deep in the body of any shark. In most sharks, this blood is carried in veins to the gills to pick up oxygen. Because the gills are in direct contact with the outside water, heat is lost, and blood is quickly cooled to outside water temperature. It is then pumped by the heart and into the arteries and back into the body. Blood returns to the gills via veins close to surface of the shark, again losing heat to the outside. A warming and cooling cycle continues, with the net result being that blood is maintained at about the same temperature as that of the surrounding water. This is also the situation in most bony fishes except tunas and billfishes.

But lamnids have a circulatory system different from most other sharks. Both arteries and veins lie deep in the body and in close contact with one another. This allows heat to be transferred from the veins to the arteries, warming the blood coming from the heart before it reaches the muscles. This arrangement is called a counter-current exchange system, because blood vessels carrying blood flowing in opposite directions exchange heat owing to their closeness. Hence, cool arterial blood is continually warmed by warm venous blood. As a result, the body and stomach temperature of a lamnid shark can be 7°C to 10°C warmer than the outside water temperature.

How does being warm-blooded help these sharks? The answer is a matter of speculation, but a popular explanation invokes the relationship of body temperature to activity level. Lamnid and thresher sharks are among the largest active sharks in cooler oceans, where they feed on active prey. (Other large, high-latitude sharks include the somniosid sleeper sharks, whose activity levels are described by their name, and the cetorhinid Basking Shark, which feeds on slow-moving zooplankton.) White Sharks are more common in cool-temperate than tropical regions, as are the marine mammals (seals, sea lions, whales) that they prefer. Salmon Sharks (*Lamna ditropis*) are common in North Pacific waters, where they feed on salmon; Porbeagles (*Lamna nasus*) are active predators in the northern North Atlantic; and makos are worldwide predators of tunas. Warm muscles are more powerful, and higher body temperatures may allow for the faster move-

ment needed to chase down and feed on fast-swimming, warm-blooded prey such as seals, sea lions, and tuna.

What kind of teeth do sharks have?

Shark teeth are usually depicted as triangular or slender pointed structures. Although this holds for many shark species, sharks and their relatives possess an impressive diversity of tooth types, strongly influenced by prey type. In addition, teeth in the upper and lower jaw may be different, and tooth type changes with age and sex in some species (see "Do a shark's teeth change during its life?" below). Differences among species are great enough that many shark species can be identified by their teeth alone.

Fish eaters tend to have long, slender teeth. Excellent examples of these are makos (*Isurus*), Ragged Tooth/Sand Tiger Sharks, and squatinid angel sharks. This is an ideal tooth form for grabbing slippery, evasive prey such as fishes and squids, which are then often swallowed whole.

If a shark's preferred prey is likely to be too large to swallow whole, then holding and cutting equipment is needed. The easily identified notched and serrated (saw-edged) teeth of a Tiger Shark accomplish this, and Tiger Sharks have one of the most varied diets of any large shark species. Tiger Shark teeth are often described as cockscomb-shaped because of their distinct uneven, notched shape and serrated edges, similar to the comb on the head of a rooster. A Tiger Shark's bite may be more effective because the teeth in each side of both upper and lower jaws are mirror images of each other. These opposing faces with a notch may increase the shearing action of the teeth, especially when the shark swings its jaws from side to side as it bites down on its prey.

Many sharks divide the labor of holding and cutting between their lower and upper jaws. In carcharhinid sharks such as the Sandbar Shark, *Carcharhinus plumbeus*, lower-jaw teeth are more slender and pointed, whereas upper-jaw teeth are triangular with sharp serrations. As the shark's mouth is closed once or repeatedly, prey are impaled and held with the lower-jaw teeth while the upper-jaw teeth work back and forth, cutting chunks of flesh and severing bone. (Some squaliform sharks such as Kitefin Sharks, *Dalatias licha*, reverse the functions and have sharper upper teeth.) Cutting is aided by head and body shaking, twisting, and even spinning. White Sharks, which take large pieces out of even larger prey such as whales, are another species with different lower- and upper-jaw teeth, at least as adults. Piranhas, another fish that bites chunks from its prey, have triangular teeth that are remarkably similar in shape to those of many sharks.

Sharks that feed on unusual prey or in an unusual manner have un-

Mako sharks are fish-eating specialists, able to chase down and grab such elusive prey as tunas. The teeth in both jaws of this Shortfin Mako Shark, *Isurus oxyrinchus*, have the classic stabbing shape characteristic of sharks that feed primarily on slippery, fast-swimming prey such as fish and squid. Photo by Didier Descouens, Wikimedia Commons, http://en.wikipedia.org/wiki/File:Isurus_oxyrinchus_Machoire.jpg

The Kitefin Shark has very different teeth in its upper and lower jaws. The upper teeth are spiky, and the lower teeth are triangular, the reverse of the condition in most sharks with different upper- and lower-jaw teeth. The Kitefin catches fish and squid with its upper-jaw teeth but uses its lower-jaw teeth to bite chunks out of large prey, as does its close relative the Cookiecutter Shark. The lower jaw of sharks is more mobile, helping in this feeding behavior. Photo by Luca Oddone, Wikimedia Commons, http://en.wikipedia.org/wiki/File:Dalatias_licha_jaw.JPG

The Cookiecutter Shark (*Isistius brasiliensis*) is a small tropical species that parasitizes tunas, sharks, and marine mammals, gouging out plugs of flesh with specialized teeth. A White Shark photographed near Guadalupe Island, Mexico, in 2011 had a fresh Cookiecutter bite mark just behind its jaws. Photo by C. S. Johnson, from Springer and Gold (1989); used with permission of the Smithsonian Institution Press

usual teeth. In some squaloid sharks, such as the 40-cm-long (15.7-in) Cookiecutter Shark, *Isistius brasiliensis*, the lower teeth are arranged in a flat band across the lower jaw and are more bladelike than the upper-jaw teeth. Cookiecutters are somewhere between parasites and predators, taking (relatively) small pieces out of large prey. Cookiecutter Sharks get their name from the wounds they leave on tunas, dolphins, whales, and even Megamouth and White sharks and humans. These wounds are circular and craterlike and about 4 cm (1.6 in) across, an exact match to the mouth size of the Cookiecutter Shark. The shark apparently attaches to the side of its victim via suction created by its movable tongue, fleshy lips, and special throat, jaw, and gill muscles. It then sinks its hooklike upper teeth and sawlike lower teeth into the prey and twists and spins about its long axis, removing a flat, craterlike cone of tissue. Similar, larger wounds have been found on seals and beluga whales and may result from attacks by large Greenland Sharks, a squaloid with a tooth type and distribution similar to a Cookiecutter's.

Sharks and rays that feed on hard-shelled prey such as clams and snails and on large crustaceans such as lobsters and crabs have broad teeth designed more for crushing than for piercing. Instead of just the outer row contacting the prey, these crushing teeth occur in bands that bring several rows into action at a time, forming a broad crushing plate. Nurse Sharks eat both invertebrates and fishes, and their teeth are broad with small points and with several rows functioning in the bite. In many rays, the teeth are entirely flattened and arranged like paving blocks on a roadbed, providing the perfect crushing surface.

Sharks and rays that feed on hard-bodied prey have rounded, flat, or pavement-like teeth good for crushing. (*Left*) Upper and lower jaws of a Cownose Ray, *Rhinoptera bonasus*, a predator on clams, showing the flattened, pavement-like dentition typical of many rays. (*Right*) Lower jaw of the Zebra Bullhead Shark, *Heterodontus zebra*. Heterodontid sharks eat crabs, sea urchins, clams, and mussels. Their sharp anterior teeth seize items that are then passed back to the stouter, rounder posterior teeth for crushing. *Left*, photo courtesy of Samantha Mulvany, used with permission; *right*, photo courtesy of Phil Motta, used with permission

Horn Sharks (*Heterodontus*—literally, "different teeth") get their scientific name from the different teeth in different parts of their jaws. Teeth grade from sharp, cone-shaped teeth for grabbing and piercing at the front margin to broad, rounded pads for crushing and grinding towards the back, a good arrangement for impaling soft prey or picking up slippery snails and clams and passing them to the back of the jaws for crushing. Bonnethead Sharks, *Sphyrna tiburo*, small members of the hammerhead shark family Sphyrnidae, also feed on hard-bodied prey and have pointy anterior teeth for grasping and flatter posterior crushing teeth. Wolf-eels—bony fishes of the northern Pacific and Atlantic oceans—feed on crabs, sea urchins, and other hard-bodied prey and have a similar pointy-in-the-front, rounded-in-the-back dental arrangement.

A few other oddities deserve mention. Selachian saw sharks (Pristiophoridae) and batoid sawfishes (Pristidae) are thought to feed in part by slashing through schools of baitfish and then picking up the wounded. Their functional "teeth" are on the outside edges of their elongate snouts and are sharp, stout, smooth, and rounded, an effective tool for maiming fish. Their jaw teeth are much less impressive, consisting of 10 to 12 rows or bands of small rounded teeth with a slightly sharpened face on the back edge. Being bitten by a sawfish would be much less traumatic than being smacked by one.

Some of the biggest sharks have the smallest teeth, and in fact, teeth may be basically nonfunctional (vestigial) in such megasharks as the Whale Shark, the Basking Shark, and the Megamouth Shark. These sharks eat

Sharks: The Animal Answer Guide

mostly zooplankton, which they strain from the water with their gill rakers as water passes from their mouths and out their gill slits.

Chimaeras (Holocephali) feed on hard-bodied prey and have evolved an entirely different solution with their rabbit-like teeth. Holocephalan teeth are not replaced, as in sharks, but instead continue to grow to make up for the wear they experience from eating hard-bodied prey. A ratfish has three pairs of tooth plates, one pair in the lower jaw and two pairs in the upper jaw. Ratfish also have tooth plates inside on the roof of their mouth, which push against the tongue during a bite.

How many teeth does a shark have during its lifetime?

Mammals, including humans, have two sets of teeth during their lives—baby or milk teeth when young, and adult or permanent teeth later. The normal total for a person is 20 milk or primary teeth and 32 permanent teeth, for a total of 52. Sharks, in contrast, replace their teeth throughout their lives, so their teeth are categorized as either functional or replacement. Functional teeth are used daily in capturing and processing prey. They sit at the jaw margins and are exposed when a shark opens its mouth to bite something. Replacement teeth lie in rows farther back and deeper in the jaws, hardening as they develop and ready to move forward in conveyor-belt fashion as functional teeth are lost during feeding or are shed as a natural process. Shedding occurs in part because the teeth are embedded in a shark's gums rather than anchored to the jawbones as in mammals and bony fishes.

A functional tooth and the replacement teeth lined up behind it make up a row. In most sharks, only the one or two outermost teeth in a row are

Functional and replacement teeth in the lower jaw of a Bull Shark, *Carcharhinus taurus*. The teeth pointing upward are the functional teeth. At least five series of replacement teeth are visible lying against the inside of the jaw cartilage, ready to move into position if a functional tooth is shed. In the third row from the left, a replacement tooth appears to be rotating into functional position. Photo courtesy of Phil Motta; used with permission

functional, although this number can range as high as eight. The number of tooth rows differs among species: some rays have only one row in each jaw, whereas Whale Sharks have more than 300 rows of very small teeth. Most sharks have 20 to 30 tooth rows, meaning 10 to 15 functional teeth on either side of each jaw. Teeth can also serve different purposes during their "functional" life. The functional teeth of the Spotted Eagle Ray, *Aetobatus narinari*, consist of a central group of fused teeth in the upper and lower jaws that are used for crushing prey. However, as these teeth are replaced from behind, the lower-jaw teeth move forward, become shovel-like, and are used to dig up prey.

The pattern of replacement also differs among species. Some sharks, such as White Sharks and hammerheads, replace teeth individually as they are worn out or lost. Squaloid sharks such as Spiny Dogfish, Greenland, and Cookiecutter Sharks replace entire rows regardless of use; replacement teeth in interior rows continue to grow and move forward, eventually displacing or replacing functional teeth. Teeth also grow as the shark grows. Teeth in the internal rows are therefore larger than the functional teeth about to be replaced.

We know from sharks kept in captivity how often replacement occurs and therefore can estimate the number of teeth a shark may have during its lifetime. Replacement rates vary tremendously among families and species, occurring every few days to every month. Rates also differ between young and old individuals of a species, younger sharks replacing teeth more often. Seasonal differences also occur, with teeth replaced more often during summer than winter, as in Nurse Sharks, which replace teeth every 9 to 28 days in summer but every 51 to 70 days in winter. Sand Tigers (*Carcharias taurus*) replace a functional tooth about every 2 days, Lemon Sharks every 8 to 10 days, Leopard Sharks (*Triakis semifasciata*) every 9 to 12 days, Horn Sharks (*Heterodontus francisci*) every 28 days, and the Small-spotted Catshark (*Scyliorhinus canicula*) at 5-week intervals. From these numbers, it has been estimated that a shark produces about 30,000 teeth during its lifetime, although the actual number remains unknown for most species.

Do a shark's teeth change during its life?

A shark's teeth are generally appropriate for the kind of food it eats. It is not surprising then that some sharks that change their diet also change their tooth type (although Bull Sharks change from a fish-heavy diet to a diet that includes tougher foods such as marine mammals and sea turtles by developing a stronger bite rather than different teeth). A good example of a tooth changer is the White Shark, which switches from a diet of mostly fish when it is less than 3 m (10 ft) long to a diet that includes marine mam-

The teeth of young White Sharks in both upper and lower jaws are slender and sharp. Only later do the upper jaw teeth take on the classic triangular shape. Slender teeth are better suited for capturing slippery prey such as the fishes that make up the diet of young White Sharks. Photo by Gene Helfman

mals when it grows to over 3 m in length. Fish-eating sharks (makos, Sand Tigers) often have long, slender teeth for grasping, whereas sharks that cut chunks from their prey have broader, serrated-edge teeth for cutting. The teeth of White Sharks, especially in the upper jaw, reflect this dietary switch, being more slender and spiky in young sharks and more triangular in older, larger individuals. When a shark switches from low-energy fish to high-energy whales, its growth rate accelerates (see "How fast do sharks grow?" in chapter 6).

One might think that being a successful yet inefficient forager on dead whales would be worth a little extra work, even if your teeth weren't quite up to the job. Recent biomechanical studies provide additional insight. The jaw cartilage of White Sharks becomes stiffened (more mineralized) after these animals reach 3 m in body length. Before that, their jaws, not their teeth, might be damaged if they chomp down on tough, hidebound prey. Spiky teeth and a fish diet may be the way they compensate for inadequate jaw strength while youngsters (although we've recently learned that even young White Sharks will feed on marine mammals if the opportunity arises).

The lamnoid White Sharks and their relatives have another type of tooth before they are born, analogous to the baby teeth of mammals. As embryos, their teeth are peglike and blunt, without cutting edges or serrations. Most embryonic lamnoid sharks feed on eggs produced by the mother while developing in the mother's uterus. It is thought that their blunt teeth prevent them from damaging the mother's uterus while being sufficiently functional to handle the eggs the mother produces. The exception in the group is the Sand Tiger, which actually cannibalizes its siblings

in the uterus (see "Do sharks lay eggs or do they give birth to live young?" in chapter 6). Sand Tiger embryos develop sharp, recurved teeth early in their development that should be superior to blunt teeth when munching on fellow pups. Researchers dissecting recently dead, pregnant female Sand Tigers have actually been bitten while probing around inside the mother.

Some sharks and rays use teeth for purposes other than feeding. In the few that have been observed mating (see "How do sharks reproduce?" in chapter 6), the male holds onto the female with his mouth by seizing her body or fins. Male Blue Sharks, *Prionace glaucus*, have longer teeth than females, presumably as an aid in holding onto females. In some rays, tooth type differs seasonally between males and females. Male Atlantic Stingrays (*Dasyatis sabina*) have the same characteristically blunt teeth as the females outside of the breeding season, a type useful for crushing clams. During the breeding season, however, the males grow pointed cusps on their teeth that they use for courtship. These teeth are shed after the reproductive season.

A particular tooth type does not necessarily limit a shark's diet. The loose attachment of teeth to jaw can be turned to advantage. Whitespotted Bamboo Sharks (*Chiloscyllium plagiosum*) have sharp, spiky teeth appropriate for feeding on soft-bodied prey such as fish and squid. However, these sharks also eat crabs that have hard shells. When biting down on a crab, a Whitespotted Bamboo Shark allows its functional teeth to hinge backward, overlapping the replacement tooth immediately behind it. The shark now has multiple rows of depressed teeth that form a flat surface more appropriate for crushing hard prey. The teeth spring back up after a bite is taken.

What is the skeleton of a shark made of?

Sharks don't have any real bone in their bodies. Most of their skeleton is made of cartilage, some of which is hardened or has fibers running through it for strengthening. Cartilage (the stuff that stiffens our nose and ears) has a number of advantages over bone. It is lighter than bone and more flexible. Hardened cartilage, as found in a shark's vertebral column, skull, and spines, gets its stiffness from deposits of calcium phosphate crystals. Shark fins are stiff, but the stiffness comes from a horny substance and not from calcification. Shark teeth aren't bony either but are instead evolutionarily modified scales covered in hard enamel.

Does a shark have senses like ours?

Yes and no. Read further.

A single vertebra thought to be from a White Shark. Several of these large vertebrae were found on a beach at Goose Island, British Columbia, and are now on display at the Whale Interpretive Centre at Telegraph Cove on Vancouver Island, British Columbia. Photo by Gene Helfman

How good is a shark's sense of smell?

Elasmobranchs have exceedingly acute olfactory (smelling) capabilities, which they use to find prey over long distances. Many popular accounts refer to sharks as "swimming noses" because of their reliance on this sense.

Detection involves paired nostrils on the underside of the snout. The nostrils lead to the enlarged olfactory bulbs of the brain via olfactory nerves. A shark's olfactory bulb takes up 3% to 14% of total brain mass, depending on the species. White Sharks have the largest olfactory bulbs relative to brain mass of any shark, skate, or ray. Hammerheads are intermediate (7% of brain mass), and skates fall at the lower end, although they are still remarkably sensitive to odors.

Sharks are attracted to blood, as we all know, but some other chemicals are even more enticing. Laboratory tests have shown that certain amino acids, the building blocks of proteins, are more stimulating than blood. Amino acids would be released into the water when a fish or crab was injured or decomposing, or even was just excreting waste products, which happens more often than injury.

Olfactory cues can be dispersed in a number of ways. One is diffusion of odor molecules through the water. But diffusion is a slow process that may disperse odors a short distance, on the order of a few meters. Longer-distance detection requires that the odor be carried by water currents, and this is probably the major way sharks locate food using their sense of smell. Operators of shark-watching boats attract White Sharks to the boats by

Form and Function of Sharks

"chumming," which involves mashing fish parts with beef blood in a bucket of sea water and ladling it overboard. Sharks approach from downstream, often zigzagging across the chum slick as they move upcurrent. The closer to the source, the stronger the odor, so by swimming up the gradient of intensity, the shark knows it is getting closer. Sharks also swing their heads back and forth while swimming, which allows them to "sample" a wider swath of ocean and gives some directional cues as their head moves in and out of the odor field. Some sharks can also locate an odor by detecting a slight difference in odor strength in the right versus left nostril, or even by the time difference the odor molecules arrive at one versus the other nostril. Either would help tell a shark the direction from which an odor was coming.

The sensitivity of sharks to chemical substances is legendary. Comparisons are difficult to express, but statements in the scientific literature talk about responses in sharks and skates to odor concentrations as low as one part odor substance in one thousand or even ten billion parts water.

Sensing chemicals includes not just smell but also taste, and the two senses are closely related. Sharks have taste buds inside their mouths and throats, especially just behind their teeth, which would be the first place a meal would land after a bite. Some sharks, such as Nurse Sharks, have a pair of external whiskers called barbels on the underside of their snout ahead of their mouth that also detect chemical cues, as do many catfishes. Saw sharks (Pristiophoridae) have barbels projecting out from their saws, in between some of the saw teeth.

Research on olfaction in sharks has led to some intriguing, and still puzzling, findings. White Sharks possess a cranial nerve in their brains that appears to be specialized for sensing chemical signals put out by other White Sharks. Such intraspecies chemical signals are called pheromones. A pheromone detector could help a male locate a female or determine her readiness to mate, either across large distances or up close, assuming a female released a pheromone when she was receptive.

Another study showed that Oceanic Whitetip Sharks (*Carcharhinus longimanus*) are able to trap and detect surface bubbles that could carry airborne odors. By swimming near the surface, a Whitetip could therefore detect scents carried on the wind, a faster and wider means of odor dispersal than water currents. Killer whale researchers use dogs perched on the bow of a boat that can smell and react to whale poop, telling them that killer whales have been in the area recently (the researchers collect the poop to see what the whales have been eating). Perhaps some sharks also learn of prey whereabouts similarly via airborne odors. And this may be a (partial) explanation for why some sharks swim at the surface in deep water, with their dorsal fins sticking out. Swimming near the surface would also

Sharks: The Animal Answer Guide

put a shark's nose in contact with any floating, odor-releasing compound at the water's surface, especially oily or fatty particles that tend to float.

How well can sharks hear?

Sharks have good hearing, especially at the low frequencies that characterize natural sounds and vibrations. Their hearing range is from 20 to 1100 Hertz (Hz), which is about the second to the sixth octave of the musical scale. The lowest note on a tuba is 16 Hz, or close to the lowest note a shark can hear. The normal range of human speech, between 85 and 255 Hz (maximally 1050 Hz for females), falls within the hearing range of sharks.

Sharks are most sensitive to and attracted by sounds produced by their prey, especially prey that might be injured or frightened. Such sounds tend to be at the low end of the shark's hearing range, and it is not surprising that sharks respond strongest to sounds below 375 Hz. Juvenile Lemon Sharks are attracted to sounds such as a quickly accelerating fish or a fish swimming with a damaged tail, and to the drumming sound produced by an injured marine catfish. A struggling grouper produces 10- to 150-Hz sounds that can attract reef sharks from as far away as 200 m (656 ft). Prey fish crunching on food can attract Lemon Sharks 10 to 15 m (33–50 ft) away.

To hear these sounds, sharks use their inner ears and receptor cells scattered around the body. Sharks have internal ears located inside their heads on both sides rather than external ears like mammals. The inner ears are linked to the outside by a very small canal that ends on the side of the head. The body receptor cells occur between the scales, particularly under the mouth and at the base of the pectoral fins. Other receptors are concentrated in the lateral line canals on the body and head, making a shark not just a swimming nose but also a swimming ear.

The concentrations of receptor cells in shark species differ as a function of ecology. Bottom-living sharks, skates, and rays have most of their receptor cells on their backs (dorsal surfaces), whereas sharks that swim above the bottom have the cells along their sides. Chimaeras, which also swim above the bottom, have receptor cells in canals that run from their nose to the tip of the tail.

Although we can detect sounds at higher frequencies than sharks can, sharks do something we can't, at least not when we're underwater. They can tell what direction a sound came from. You can demonstrate your inability to do this by submerging yourself in a pool or pond, closing your eyes, and having a friend bang on a bottle off to one side while underwater. The noise appears to be coming from everywhere, something scuba div-

ers know from trying to locate a buddy who is banging on a scuba tank to alert other divers (perhaps to the presence of a shark in the area?). Localizing a sound requires your brain to be able to detect the slight difference in the sound's arrival time at either ear. Because human hearing evolved on land, we have little trouble localizing terrestrial sounds, which travel at 340 m (1,115 ft) per second. Sound travels faster—in fact, four times faster—in water. Nothing in our evolutionary history prepares us for that kind of speed. Sharks are exquisitely adapted to the aquatic environment; we are not.

How well can sharks see?

Although sharks live and can feed in the dark or in very turbid water, they also have good vision because of well-developed eyes and large optic lobes, the part of the brain devoted to vision. Sharks tend to be a little farsighted, meaning they are able to focus best on objects more than 23 cm (9 in) away. But sharks don't have to read textbooks, which is good because their overhanging snouts would block their view of objects that close. For locating objects close to their mouths, they probably rely more on electrical input (see "How do sharks detect electric fields?" below).

A shark's eyes are built on the same plan as the eyes of all vertebrates. Moving along the path of light, the eye consists of an outer, clear cornea; a lens (spherical rather than disk-shaped as in bony fishes and mammals); a slit-like pupil surrounded by an iris diaphragm; and a retina made up of receptor cells called rods and cones. Light enters through the cornea. Under differing conditions of brightness, the iris squeezes closed or opens up to regulate the amount of light entering, something bony fishes can't do. The lens is moved back and forth for focusing. Cone cells in the retina operate best under daytime, bright light conditions and provide acute vision (high resolving power for smaller objects moving short distances). Cone cells are also tuned to differing wavelengths of light and are responsible for color vision in those species that can discriminate color (see "Can sharks see color?" below). Rod cells are more sensitive to light and operate chiefly under low light conditions such as twilight and nighttime, providing a black, white, and gray view of objects and their movements.

Many sharks can rotate their eyeballs to follow an object, another thing bony fishes can't do. Some sharks and many rays have eyelids that protect the eye from physical harm or block out bright light. In some sharks such as the evolutionarily advanced carcharhinid reef sharks, the eyelid is an opaque tissue called a nictitating membrane that is pulled over the eye to protect it. White Sharks lack any kind of eyelid but instead rotate the

eyeball backward when attacking active prey. This moves the more delicate parts of the eye out of harm's way, exposing the tougher back of the eyeball, called the sclera, to flailing seals or surfers.

An additional feature behind the retina is a layer of reflective crystals called the tapetum lucidum (literally, "bright carpet"). The tapetum is responsible for the "eye glow" that occurs when a light is shone at a shark, a cat, or a crocodile at night. Many nocturnally active animals have a tapetum, which acts as a light amplifier. Light rays pass through the retina, activating the receptor cells. The tapetum reflects the light that wasn't captured in the retina back through the retina, stimulating additional rod cells on the way back out. (The eye shine we see is the tapetum-reflected light that wasn't captured moving in either direction.) A tapetum therefore maximizes the amount of light captured. As a result, sharks are about 10 times more sensitive to light than humans, and even twice as sensitive as a cat.

Sharks tend to have larger eyes than skates and rays, and sharks that live in the open ocean closer to the surface have larger eyes relative to body size than sharks that live in the deeper, darker waters, where vision would be expected to be less important because of the comparative lack of light. Sharks that feed on active prey (fishes, squids) also tend to have larger eyes than more sluggish benthic feeders that prey on less mobile prey such as crabs or clams. Chimaeras have even larger eyes relative to their body size than sharks, skates, or rays. Their large, bulging eyes (and buck teeth) are what gave them the alternative name of rabbitfish.

Can sharks see color?

The question of whether sharks see color has intrigued scientists for decades. Testing for color vision is more difficult than it would seem. Some animals that have visual structures that should allow them to see color—namely, the right kinds of cells in their retinas—do not respond behaviorally to different colors (cats are an example). Sharks respond to different colors, but the difference might be to brightness and contrast rather than to color.

Recent advances in microscope types appear to have finally answered the question: sharks cannot see colors, but rays and chimaeras can. Tests on the retinal cells of 17 shark species, including Tiger and Bull sharks, indicate they lack the necessary cell types. These species possess cones, but to see color an animal needs more than one cone type, with each type most sensitive to different wavelengths of the visible light spectrum. The sharks tested either possessed only rods or had only one cone type. This means they can't see color but instead are sensitive to brightness and con-

trast. Shark coloration—or rather, the lack of it beyond bold darks and lights—supports this conclusion. Skates also apparently have only rods in their retinas.

Rays and chimaeras are different from sharks. Some have three cone types, suggesting they can discriminate among colors. This makes sense given that some rays have colors such as blue spots and live in shallow water where all wavelengths of light penetrate. Freshwater stingrays and the Bluespotted Stingray, *Dasyatis kuhlii*, are striking but not the only examples. The Bluespotted Stingray has the necessary three types of cones in its retina that should enable it to see colors, as does the Yellow Stingray, *Urobatis jamaiciensis*. At least one chimaera, the Australian Ghost Shark, *Callorhinchus milii*, has three cone visual pigments and therefore could have color vision. Some chimaeras (such as the Pointy-nosed Blue Chimaera, *Hydrolagus trolli*) are reddish or purplish, but red light does not penetrate to the depths where most chimaeras normally live, which means their bodies would appear black. But even some of these deep-living species migrate to shallow areas to spawn, so perhaps color vision comes into play at that time. Once again, chimaera biology is full of mysteries.

What about sharks' seeming preference for yellow? Researchers attached baits to different colored floats. They found that sharks preferred baits tied to international orange floats. Life jackets are often this color because *we* see it well. Combining life jacket color with shark feeding preferences has prompted Navy experimenters to call the color "yum-yum yellow." (See the MythBusters episode "Sharks Prefer Yellow?" MythBusters gave the yum-yum yellow idea "plausible" status.) But this attraction could be to brightness rather than the color itself.

Although the new research on shark color vision looked at only 17 of the 400 or so shark species, it strongly suggests that sharks are color blind. A shark may be able to smell small amounts of blood in the water, but it doesn't know that it's red.

How do sharks detect electric fields?

In addition to the conventional five senses of touch, taste, smell, vision, and hearing, sharks have a sixth sense that we (and all mammals except the duckbill platypus) lack. Sharks, rays, and chimaeras can detect weak electric fields, or at least they all have the Ampullae of Lorenzini that are the electrosensory structures responsible for detecting such stimuli in species known to detect weak electricity.

An ability to detect electrical fields aids sharks in finding nearby food, and in fact when a shark finally homes in on a food source, information from the electrical output of the prey may be more important than its

The small holes on the head of this Blue Shark are Ampullae of Lorenzini, the input structures for a shark's electrical sense. These pores were discovered by Italian anatomist Stefano Lorenzini in 1678. Lorenzini had no idea of their function; that understanding took another 300 years. Photo by Austin Gallagher, http://aust ingallagher.com; used with permission

smell, appearance, or sounds. This has been demonstrated in the field and laboratory with sharks that were attracted to baits suspended next to active electrodes. The sharks were clearly first lured in by the smell and tended to approach the odor source. But at the last moment, they ignored the bait and attacked the electrode. In fact, ampullary electroreceptors are located around the mouth of many elasmobranchs—which is to say, in their blind spot so far as vision is concerned. This positioning could allow a shark or ray to home in precisely on a potential food source solely by electroreception, effectively lining the food up in its "electrical sights" and then engulfing the prey.

How fine-tuned is this electric sense? We can only detect electrical fields of one-tenth of a volt (0.1 V) or stronger. Sharks can detect fields of 10 billionths of a volt (0.00000001 V), which is ten times what is needed to sense the electrical output from a prey fish. It's difficult to put this in a meaningful context. But such sensitivity would be sufficient, at least theoretically, to allow a shark to detect the electrical output of a 1.5-V flashlight battery over a distance of 1,300 km (800 miles), or roughly the distance between Seattle and San Francisco (or New York City and Jacksonville, Florida). More meaningful to us, a shark could theoretically sense the bioelectric output of a swimmer 1 to 2 m (3–6 ft) away.

Such sensitivity means sharks can use electrical output to find prey. This is well known from lab experiments with species such as the Atlantic Stingray and two scyliorhinids, the Swell Shark (*Cephaloscyllium ventriosum*) and Small-spotted Catshark. These animals can find and capture prey fish buried in the sand under conditions that eliminate all other sensory cues. Sharks can also sit motionless in the dark and snap up prey that swim nearby by using electrical signals. Electrical cues are also used in social en-

counters. Male Round Stingrays (*Urolophus halleri*) find females buried in the sand by detecting the electrical output produced by the female's breathing activity. In electrogenic skates (those that have special tail organs for producing electricity), the output is used during social interactions.

Theoretically, sharks are sensitive enough that they could detect the earth's magnetic field and determine their compass headings when crossing oceans. It is thought that sharks such as hammerheads navigate to and from particular areas on a daily basis by homing in on electrical properties of the seafloor, including variation in the earth's magnetic field. Experimental evidence supports this idea. Stingrays in lab pools can learn to orient to food in fields weaker than the earth's field. When experimenters reverse the field polarity, the stingrays reverse the location where they search for the food, a behavior which suggests that geomagnetic cues can be used in normal daily activities. Elasmobranchs are thought to have small particles of magnetite (an iron substance) in their inner ears that could be used in geomagnetic orientation. When magnets are placed over the predicted region of magnetite in the head, rays lose their ability to orient to a magnetic field.

In addition to detecting electricity, some rays (but not selachian sharks or holocephalan chimaeras) produce unusual amounts of electricity. Best known and understood is the strong electric output of the torpedo rays (family Torpedinidae). Also called numbfish, torpedo rays produce electrical discharges of up to 50 V and 50 A, producing an output approaching 1 kW. Muscle contractions in all vertebrates involve some electrical production; torpedo rays (and electric eels and their relatives) have evolved muscle systems that don't contract but are wired together to maximize this electric output. In the case of the torpedo rays, the muscles around their gills and pectoral fins are arranged and modified to produce strong discharges. With the pectoral area given over to electricity production, torpedo rays are among the few raylike fishes that swim with their tail fins instead of their pectoral fins.

Torpedo rays can use electric discharges to catch prey. At night, torpedos swim slowly above the bottom; when they encounter a small fish, they "embrace" it with their pectoral fins and discharge electricity, stunning the prey and making it easy to capture. This was first discovered by research divers who placed prey fish on the ends of poles that were wired to flash bulbs. When the ray embraced the prey, the flashbulb fired.

Electric output is used in defense as well as offense. A captured or speared torpedo ray will shock you when you handle it. Unfortunate divers who bump into a torpedo at night may fall victim to this defensive response. Mysterious instances of drownings have occurred among scuba divers in southern California. The victims were night diving alone and were found on the bottom with plenty of air in their tanks and no visible signs of injury.

It has been speculated that such divers bumped into a torpedo ray, received a strong shock, became disabled, and inhaled water.

Some skates (Rajidae) emit a weak electric discharge from specialized organs in their long tails. The output differs among species and may help identify conspecifics (members of the same species) during mating. Weak electric discharges are unknown among any other marine fishes, bony or cartilaginous.

Do sharks sleep?

We really don't know whether sharks sleep. Sleep in other vertebrates, including bony fishes, involves a reduction in activity and metabolism, and a decreased sensitivity to external stimuli. Sharks that lie on the bottom (see "Must a shark keep swimming to breathe?" below) are less active but are still consuming energy. In fact, juvenile Lemon Sharks that lie on the bottom and must pump water over their gills to breathe use about 10% more energy than does a juvenile swimming slowly with water passing over its gills simply because of its forward movement. However, Lemon Sharks are usually active, unlike several species that do "rest" on the bottom for extended periods (but still actively pump water in their mouths and out their gills). This "resting" group includes Nurse, wobbegong, and Whitetip Reef (*Triaenodon obesus*) sharks as well as angel sharks and many batoid skates and stingrays.

The idea of sleeping sharks gained credibility with the discovery of normally active sharks lying on the bottom in caves off the Yucatan (Caribbean) coast of Mexico. These "sleeping sharks of the Yucatan" were mainly Caribbean Reef Sharks (*Carcharhinus perezi*). Over the years, another four species of sharks have been seen at this cave locale, including Tiger, Lemon, and Bull. Low-salinity water flows out of the caves, and the sharks are remarkably quiet and approachable. Why they are there and whether they are sleeping remains a matter of conjecture.

Do sharks have scales like other fishes?

Sharks have scales but they are very different from the familiar scales of bony fishes. Bony fish scales are flat, circular, relatively smooth, and overlapping. Shark scales, called placoid denticles, have ridges and valleys and bumps, vary greatly in shape among species, and seldom overlap. They can be close together or scattered almost randomly around the body. Seen from the side, many shark scales have one or several pedestals, often with a flattened top and sometimes with a recurved hook facing backward. Form follows function: fast swimming sharks have scales that, because of

the ridges and flat-topped pedestal, decrease turbulence around the body and thus serve as streamlining. In fact, the keels of high-performance sailboats mimic this design in their "winged keels," a design adaptation that led to Australia's winning the America's Cup sailboat races in 1983. (The Ozzies' winged keel innovation was a closely guarded secret.) Swimsuits designed for competitive swimmers, such as the Speedo Fastskin, have built-in ridges emulating a shark's denticles to reduce drag across a swimmer's body. Critics claim that these swimsuits have contributed to record-breaking swim times, and such "performance-enhancing" suits have been banned from some competitions.

The reduced turbulence created by a shark's denticles also allows a shark to swim through the water quietly, a definite advantage for a predator trying to sneak up on its prey. Hydrophones (underwater microphones) detect much less noise from swimming sharks than from swimming bony fishes. Not surprisingly, the scales of bottom-dwelling and slow-swimming sharks (Basking Shark; Bramble Shark, *Echinorhinus brucus*; Thorny Skate, *Amblyraja radiata*), as well as many rays, lack apparent streamlining features and are instead enlarged for protection.

The roughness and toughness of shark scales have long been exploited by humans. Dried shark skin was used as sandpaper before modern technology perfected the human-produced variety. Swimmers who get bumped by sharks are often scraped and even cut because of the roughness of a shark's body.

Scales appear on different parts of a shark's body in modified forms. You could say that the scales extend along the body and into the mouth because shark teeth are actually modified scales, growing from the same embryological structures. Another evolutionary modification of scales is the barb of a stingray, which is a modified, elongated placoid scale with sawtooth edges and a venom gland at its base.

Although scales and teeth arise from the same embryonic structures, sharks do not shed old scales and grow new ones the way they grow new teeth. However, if a shark loses a scale because of an injury, it can replace it. New scales are obvious, having a clear appearance rather than the opaque look of older scales.

How do sharks breathe under water?

Most selachian sharks breathe passively: as they swim, water enters their mouth and passes over their gills, exiting through the five (or six or seven) gill slits. This kind of breathing is often called "ram ventilation." Oxygen is taken in and carbon dioxide released at the thin membranes of the gill filaments, where blood vessels are a few cell layers away from the outside

water. Chimaeras also swim actively most of the time and breathe in this manner.

Sharks that lie on the bottom (Nurse, wobbegongs, Whitetip Reef), many skates (including sawfishes), and rays must actively pump water into their mouths and out their gills. In many such benthic (bottom-living) elasmobranchs, water is also sucked in via the spiracles, which are two holes on the top surface of the head that are derived from structures associated with the first functional gill arch.

Must a shark keep swimming to breathe?

Sharks that stay up in the water all the time must keep swimming not only to breathe but also to keep from sinking. Details on breathing are given in the answer above. But in many sharks the tight link between swimming and breathing goes a step further. In sharks that must keep swimming to breathe, the contraction of their swimming muscles also helps keep blood flowing through their arteries and veins. These sharks have relatively small hearts, or at least hearts too small to pump sufficient oxygen-carrying blood without the added "muscle pump" provided by swimming.

How do sharks swim?

The manner in which sharks and their relatives move through the dense medium of water is a fascinating topic worthy of some detailed exploration. Most sharks swim by passing S-waves of muscle contraction down their flexible bodies, starting at the head, increasing in amplitude (side-to-side extent), and ending at the tail. This snakelike movement of waves along the body limits how fast most sharks can go. A few of the faster sharks—such as the lamnid White, Salmon, Porbeagle, and makos—keep the front part of their bodies rigid and sweep their tails back and forth rapidly. The fastest shark is the Shortfin Mako, which can swim 32 kph (20 mph). Makos feed on fast-swimming fishes such as tunas that swim in a similar fashion. Wind-up bathtub shark toys and the helium-filled, radio-controlled "air swimmers" (www.airswimmers.com), with their stiff bodies and flapping tails, actually mimic this type of swimming quite well.

A great deal of effort, and no small amount of controversy, has gone into the study of shark swimming. Sharks lack the gas-filled swim bladders that make bony fishes neutrally buoyant in the water column, allowing bony fishes to maintain depth without sinking or rising. Sharks instead have oily livers that provide buoyancy, but the buoyancy provided by these organs is not enough to make up for all the weight of the muscle and cartilage. As a result, sharks tend to sink, albeit slowly, if they aren't moving

forward. One possible exception is the Sand Tiger shark, which is known to gulp air at the surface and hold it in its stomach, providing buoyancy and allowing it to hover motionless in the water column. This has been observed in public aquariums, where this species is often kept, and even in the wild. Sand Tigers in captivity have also been known to expel bubbles out their cloaca. Shark farts?

Forward swimming (no selachian shark can swim backwards) keeps sharks from sinking because of the lift provided by their fairly rigid, out-stretched pectoral fins, much in the way the fixed wings of an airplane provide lift. But the larger upper lobe of the typical tail fin would theoretically force the nose of the shark down, because the body would tend to pivot around the pectoral fins. As a result, sharks would have to brake with their pectoral fins to counter the tendency to nosedive. This would be a very inefficient and energetically costly way to move through the water, and everything we know about sharks points out that they use and conserve energy efficiently.

The answer provided by video analysis suggests that the upper and lower lobes of the tail produce mostly *forward* thrust, any pivoting overcome by the lift provided by the pectoral fins and flat underside of the head. These forces are equal and opposite to the weight of the shark in the water. Braking by the pectoral fins is unnecessary, although climbing, diving, and turning are accomplished by adjustments to the pectoral fins, which are more flexible than fixed, horizontal airplane wings.

Although most of the energy for swimming comes from muscular contraction, sharks are exceptional among fishes because their skin also aids in propulsion. Shark skin includes inner layers of collagen fibers that act as if they were tendons. Different fiber layers are oriented at angles to one another, making a shark's body a cylinder reinforced with wound fibers. Such a cylinder is strong and bendable but not compressible. Pressure builds up inside the cylinder, making the shark's body a pressurized cylinder with an elastic covering. As a shark swims faster, it basically pumps more pressure into the cylinder with each tail stroke; the pressure during fast swimming is about 10 times what it is during slow swimming. The higher the internal pressure, the stiffer the skin becomes, increasing the energy stored in stretched skin.

Body muscles are attached not just to the vertebral column but also to the inside of the skin. (For this reason it is very hard to "skin" a shark.) As muscles on one side of the body contract, muscles and skin on the other side are stretched. The stretched skin is very elastic, but stretched muscle is less so. As muscles on, say, the right side relax, the energy stored in the skin on the left side is released. In this manner, the shark's skin helps start

the pull of the tail across the midline, increasing the power output at the beginning of each propulsive stroke.

Relying on an oil-filled liver rather than a gas-filled bladder might seem an inferior solution to the buoyancy challenge, but oil has distinct advantages over gas. Bony fishes control their buoyancy by filling or emptying their swim bladder as they move up and down in the water. Gas secretion and absorption take time, and this limits how fast a bass or a perch can move up and down. Oil, unlike gas, is incompressible and provides constant buoyancy regardless of depth and pressure. Sharks can therefore move up and down rapidly and repeatedly over tens or hundreds of meters without having to make adjustments in their buoyancy control mechanism. Vertical movements of 500 to 1,000 m (1,600–3,200 ft) are known in hammerhead, thresher, Basking, Whale, and White sharks that have been followed using telemetry tags. Few if any bony fishes move up and down as much. Sharks gain at least one additional advantage from their oil-filled livers. They can use the oil as an energy reserve, metabolizing it when they are starving.

Some sharks have huge livers that make up as much as 30% of their body weight. Deep-sea squaloid sharks that live under tremendous water pressure, and some open-water species such as Whale, Basking, Tiger, and White sharks have large livers and are almost neutrally buoyant. A Basking Shark that weighs 1,000 kg (2,200 lb) on land weighs only 3.3 kg (7.25 lb) in water, or about 0.3% of its weight on land, owing to the buoyancy provided by its large, oil-filled liver. Remove the liver from a dead shark, and the shark sinks rapidly. But so much oil in one place subjected Basking Sharks to intense hunting, and they were speared out of existence in many places. The species is now considered endangered.

The batoid skates and rays have mostly lost their tail fins over evolutionary time, so they rely more on various movements of their pectoral fins to propel them. (Torpedo rays are an exception. They have modified their pectoral fins to produce strong electric pulses and move around using their tail fins.) A typical ray swims by passing waves of muscle contraction down both pectoral fins, front to back, rather than by flapping the fins. Myliobatid rays spend more time in the water column than most other rays, either to move from place to place (Spotted Eagle Rays) or to feed (plankton-eating manta rays). Myliobatids flap their pectorals up and down. The animation of Mr. Ray in *Finding Nemo* was a very accurate portrayal of the graceful way in which a Spotted Eagle Ray swims.

Watching a manta ray "fly" through the water, diving and turning with ease and grace is a beautiful sight. Engineers have built robots that attempt to mimic the flapping flight and maneuverability of manta rays in an effort to design more efficient autonomous (self-propelled) underwater ve-

hicles, or AUVs. The mantas are still better, able to turn around quickly in about half a body width, while the best AUV needs to move forward 1.3 body widths to complete a moving turn. The research involved in developing these AUVs has led to several discoveries about manta rays, including what actually moves a manta through the water. Analysis of videos shows that rippling waves of muscle contractions pass from the front to the back of a manta's pectoral fins and from the base of the fin to its tip as the fins move up and down. (Recall that most rays move via such undulations passing back along their pectorals, the pectorals remaining relatively horizontal.) The front-to-back waves of the manta fins actually provide four times more forward propulsion than the up-and-down flapping. "Mantabots" so far have managed to duplicate only the flapping parts of the movement.

Chimaeras swim primarily by flapping their pectoral fins up and down. For forward movement to occur, the angle of attack of the fins would have to change as the fin moves; otherwise, a ratfish would just bounce up and down in the water. Available studies suggest just such a wave passes along the pectoral fin with each flap, producing forward thrust. Movement is made possible by muscles that extend farther out on the fin than in sharks such as the Spiny Dogfish that don't flap their pectorals. The ratfish downstroke is stronger than the upstroke, perhaps to provide some lift and keep the negatively buoyant ratfish from sinking. Unlike aquatic birds (such as penguins) and fishes (coral reef wrasses, kelp bed surfperch) that also flap their pectorals when swimming, the ratfish moves the tip of its pectoral fin in a vertical oval with each stroke rather than a figure eight. Chimaeras once again do things differently from other animals, and much remains to be learned about their swimming.

Some sharks and skates don't just swim: they also walk along the bottom. Little Skates (*Leucoraja erinacea*) move across the sea floor in a curious manner called "punting," named after the way that people move a small boat across shallows by pushing against the bottom with a pole. The front part of a Little Skate's pelvic fin has leglike parts with flexible joints. A Little Skate pushes off the bottom with both "legs" at the same time and then glides on its outstretched pectorals. The skates can turn by putting down one leg and pivoting on it (see http://faculty.une.edu/com/dkoester/punting). The Epaulette Shark, *Hemiscyllium ocellatum*, also walks along the bottom, pushing off with the front edges of its paired fins. It pushes with its left pelvic and right pectoral simultaneously, then its right pelvic and left pectoral, bending its body and moving very much like a giant salamander. The cartilages in its paired fins have jointlike gaps that allow the fins to rotate, making them more effective walking sticks.

Many high-quality videos of swimming elasmobranchs can be found by searching the ARKive website, at www.arkive.org.

How do a shark's fins function during swimming?

While pectoral fins are providing lift and some turning ability as explained above, the propulsive force for swimming comes from the dorsal fins and tail. Most sharks have two dorsal fins, the first usually larger than the second, separated by a considerable gap. The upper lobe of the asymmetric tail can be thought of as a third fin in line with the dorsal fins, again separated from the second fin by a considerable gap. The distances between the three fins are apparently determined by the size of the fins, their shapes, and the form of the muscular waves passing from head to tail of the swimming shark. Each fin tapers posteriorly, leaving behind it a wake as it moves through the water. This wake is displaced laterally by the waves passing down the fish, so the wake itself follows a snaking path that moves posteriorly as the fish moves through the water.

Calculations of the timing differences and wave nature of the wake suggest an ideal distance between fins that would maximize the thrust of the second dorsal and particularly of the tail. If timed correctly, the trailing fins can push against water coming toward them from the leading fins. Such an interaction between flows would increase the thrust produced by each trailing fin. Measurements of swimming motions and fin spacing in six species of sharks indicate that this is exactly what happens. Unlike bony fishes, which use their foldable dorsal fins primarily to speed up or brake, or fold them to reduce drag while cruising, sharks use their dorsal and tail fins as additional interacting thrusters.

Sharks have taken the relatively inefficient eel-like swimming mode imposed by their flexible bodies and combined elastic skin, their rigid but carefully spaced median fins, and an asymmetric tail that produces a constant direction of thrust to achieve an efficient compromise between cruising, acceleration, and maneuverability.

How long can a shark live out of water?

An interesting question, one we hope few people will try to answer experimentally. The shark literature contains statements such as "There appear to be few examples of hypoxia-tolerant elasmobranchs," meaning few sharks can tolerate low oxygen conditions. Available evidence suggests that the few sharks best able to survive periods of exposure in air are those that live in shallow areas such as tidepools, where they may get stranded when the tide goes out. These include the Brownbanded Bamboo Shark (*Chiloscyllium punctatum*) and the Blind Shark (*Brachaelurus waddi*), both of which are reported to survive 12 to 18 hours of air exposure. Blind Sharks get their name from their habit of closing their thick lower eyelids when in

air, a response that presumably prevents their eyes from drying. Epaulette Sharks, a shark that feeds at night over shallow tidal flats, can live for at least an hour without oxygen at water temperatures as high as 30°C (86°F) (see "How do sharks survive droughts?" in chapter 5). Researchers working with live sharks in tag-and-release programs try to get sharks back into the water in less than four minutes; they also keep sea water from a hose running across a shark's gills during the tagging operation.

Why do hammerhead sharks have such strange heads?

Over the years, a number of hypotheses have been put forward to explain the function of the "expanded rostral cartilage," or "cephalofoil," of hammerhead sharks. The head cartilage of hammerheads is greatly expanded from side to side, giving it the hammer or mallet shape from which the group gets its common name. But people often forget that this widening is also accompanied by flattening of the entire structure. When viewed from the front, a hammerhead's head looks like someone smashed it with a hammer, with the upper surface rounded and the lower surface flat. From the front, the head looks like a wing, or airfoil, which is where the "cephalofoil" ("head wing") name comes from.

Ideas about the function(s) of this greatly modified (for a shark) shape include better binocular vision, a wider path to detect odors and locate them ("stereo-olfaction"), improved electrical sensitivity because the receiving poles are wider apart, and better hearing because the lateral line receptors are spread across the head. The broader head could also have a hydrodynamic function, providing additional lift to the front end to counteract downward pivoting caused by the asymmetric tail, or better maneuverability when the head serves as a forward-end rudder. Or it could somehow improve prey catching and handling.

An important fact to remember in this search for adaptations is that the family of hammerheads, the Sphyrnidae, contains eight species that vary substantially in the shape and "spread" of the hammer. Bonnetheads have a slightly broadened head, only about 1.5 times wider than long. (Head length is measured from the tip of the snout to the back of the mouth.) At the other extreme is the Winghead Shark (*Eusphyra blochii*), in which the head is short but exceedingly wide, almost 5 times wider than long. (Don't think that evolution has produced progressively wider and wider heads; the Winghead is evolutionarily the most primitive member of the family and the Bonnethead the most recently evolved.) Most of the other hammerhead species have heads about 2 to 2.5 times wider than long; and "normal" carcharhinid sharks, the group from which the sphyrnids evolved, have heads about as wide as they are long. So any presumed, adaptive function of

The hammer, or cephalofoil, of a Great Hammerhead Shark. Several explanations have been offered for the hammerheads' expanded head. **More than one may be correct.** Photo by Gene Helfman

the hammer would have to account for this variation while ignoring other variations in the shape of the head. Also, nothing says the head has to serve only one purpose, or that the head works in the same way for all members of the family.

Experimental evidence now shows improvement in several senses that can be attributed to the expanded head. Sphyrnids have greater binocular vision than pointy-nose sharks such as Lemon and Blacknose (*Carcharhinus acronotus*). Binocular vision is the overlap in the fields of view of the two eyes and is what gives us depth perception; in humans, the fields overlap 120 degrees (our entire field of view is 190 degrees, 120 of which are seen by both eyes). The binocular overlap of Bonnetheads is 13 degrees, for Scalloped Hammerheads (*Sphyrna lewini*) it is 32 degrees, and for Wingheads it is 48 degrees, corresponding to the differences in relative head width. Lemon and Blacknose sharks have only about a 10-degree binocular overlap. The overlap increases as a shark moves its eyeballs and swings its head back and forth, improving by 37 to 39 degrees for Scalloped Hammerheads and Bonnetheads. Having eyes out on stalks also allows hammerheads to see behind them better than nonhammered species. Combined with the forward binocular overlap, a view behind means these sharks have a full 360-degree spherical picture of their world, front to back and top to bottom. So a visual advantage does exist.

Although hammerheads are not more sensitive to odors than other sharks, they may have increased stereo-olfaction. Anatomical differences among hammerhead species show that hammerheads with broader heads can distinguish between odors striking their different nostrils better than more narrow-headed species. Additional stereo-olfactory ability could

Form and Function of Sharks

come from the lag in time when odors strike the right versus the left nostril. Smooth Dogfish (*Mustelus canis*) respond to this time lag, turning toward the side where the odor first strikes. Having the nostrils out on the ends of the hammer means this time lag is greater and allows a hammerhead to sense the direction from which an odor is coming over a greater encounter angle than narrow-headed species.

The jury is still out on the possibility of improved electrical sensitivity. Hammerheads, especially juveniles, have more electrosensitive pores (Ampullae of Lorenzini) on the undersides of their heads than other sharks, an increase made possible by the greater surface area of an expanded head. But hammerheads are not necessarily more sensitive to electrical stimuli than other sharks. Still, they might be able to "sweep" a larger area electrically than a pointy-headed shark.

Tests of improved hydrodynamic efficiency have proved surprising. The idea that the head could serve as a front-end rudder that allowed greater maneuverability provided results opposite to expectation. Hammerheads are more maneuverable, meaning that they can make faster turns than carcharhinid sharks. But the head isn't used as a rudder but more as a stabilizer; a hammerhead's body doesn't roll as much in a sharp turn, and the head may help it keep level. It's just such unanticipated results that make research exciting for shark scientists.

Having a broad head does allow hammerheads to catch their favorite prey, stingrays. Divers have seen hammerheads pinning stingrays to the bottom, which would be harder to do with a pointy head. And some go even further, using their heads literally as a hammer. Great Hammerhead Sharks have been observed using the flat underside of their head to deliver powerful downward blows that knock a stingray to the bottom before pinning it down, pivoting around, and biting pieces off the ray's wings.

Using the expanded head to hammer prey is a good example of the way evolution capitalizes on innovation. The expanded hammer probably evolved first because it improved vision or olfaction, and only later was it put to use to strike and pin prey. Often, when a species evolves a new structure that works well for certain purposes, the same structure gets put to use in novel ways that probably had little to do with its initial evolution. For example, the saw of a sawfish probably evolved as a means of striking baitfish up in the water column, disabling them, and making them easier to capture (also a major use of the sword of Swordfish, sailfish, and marlins). By chance, the same structure in sawfishes is also useful for digging in the mud and dislodging organisms there, a very different use of the saw.

But another "rule" of evolution is that every adaptation carries some liability or trade-off. Such is the case in hammerheads. As with all carcharhiniform sharks, hammerheads are live-bearers, with the young emerging

from the mother tail-first. However, an expanded rostral cartilage would easily hang up in the mother's birth canal. To facilitate passage, the head cartilages of newborn hammerhead sharks are soft and pliable at birth and fold up as they pass outside. They then stiffen and straighten shortly after.

What kind of brains do sharks have?

Receiving sensory information, interpreting it, and acting on it through muscular activity and behavior requires brain power. Sharks, in contrast to their media image as mindless killers, have relatively large and complex brains.

Unlike a mammal's brain such as ours, which is folded back on itself, shark brains are arranged in a line, divided into hindbrain, midbrain, and forebrain. The spinal cord enters through the skull and connects with the hindbrain, which is then divided into brainstem and cerebellum. Cranial nerves associated with hearing, balance, and electric sense enter the brainstem. The cerebellum sits on top of the brainstem and coordinates muscular activity.

The top of the midbrain has two swellings, the optic lobes. These are relatively small and extend out via the optic nerves and eyeballs and regulate vision. The forebrain consists of olfactory organs associated with smell and are relatively large. The front part of the forebrain is the cerebrum, which deals with learning and memory. The forebrain also houses the hypothalamus and pituitary gland. The hypothalamus produces chemicals (hormones) that affect heart rate, thermoregulation, metabolic rate, salt and water balance, food intake, and digestion. It also houses the epiphysis, the organ that senses day length and perhaps sun position. The pituitary secretes a number of hormones, including those that regulate blood pressure and the concentration of various chemicals in the blood associated with feeding, growth, kidney function, calcium levels, metabolic rate, and sexual activity.

How large is a shark's brain?

Researchers describe shark brain size in terms of the ratio of brain weight to body weight. This is thought to be a better measure for comparing animals than absolute size because bigger animals will tend to have larger brains. Squatinid angel sharks and cetorhinid Basking Sharks have the smallest brains relative to body weight. Squalid dogfishes and rajid skates also have relatively small brains. White Sharks have larger brain-to-body weight ratios than dogfishes or skates, although a 3.7-m (12-ft) White Shark's brain weighs only 35 g (1.2 oz), or 0.008% of its total body weight,

whereas our brains account for about 1.9% of our body weight. This gives us about 40 times more brain mass in absolute terms and 240 times more brain relative to body weight than White Sharks. Be proud.

Carcharhinid requiem sharks and dasyatid stingrays have relatively larger brains, and odontaspid Sand Tigers have even larger brains. Mobuline manta and devil rays have the largest brains relative to body size among elasmobranchs. In fact, the mobuline rays have larger relative brain sizes than most bony fishes, and their brains are even larger than those of some birds and mammals. (A different measure, called the encephalization quotient, places White Sharks fairly low on the intelligence scale and thresher sharks and Nurse Sharks near the top.)

Chapter 3

Shark Colors

Why are so many sharks dark on top and light on the bottom?

Most animals that live up in the water have dark tops and light bottoms. This is true not just of sharks and rays but also of dolphins, killer whales, penguins, and even crocodiles. The coloration is known as "countershading." Countershading is adaptive because it makes the animal blend into the open-water background against which it is seen by predators and prey alike. Blue Sharks and makos, which are open-ocean species, are also countershaded but tend to a dark bluish back color that is probably a match to the indigo blue background of the open seas.

A countershaded shark usually has a dark back that grades to a light belly, although in many sharks the transition from dark to light is abrupt. Regardless, the principle in operation is that the shark reflects or absorbs light that makes it blend into the background because dark colors absorb incoming light and light colors reflect it. A countershaded shark disappears because its color is opposite to the distribution of light in water, creating a target that is identical to background. The fish therefore reflects or absorbs light that is roughly equivalent to the background against which it is seen at all viewing angles, dark against dark, light against light, intermediate against intermediate.

Light coming from above is called downwelling, from the side sidewelling or spacelight, and from below upwelling. Countershading is easiest to understand when viewing a shark from above. A shark with a dark back absorbs very bright downwelling light, creating a dark target against the dark background of dim upwelling light. However, most fishes are viewed

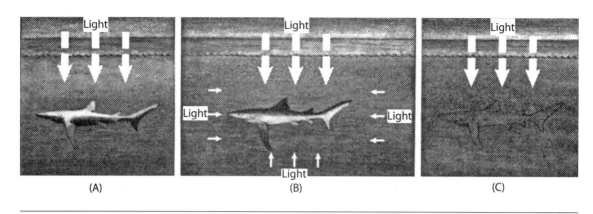

Most sharks are countershaded: darker on top, gradually lighter on their sides, and brightest on their bellies. This color pattern makes them disappear in the water column because their colors reflect light that matches natural background light. (A) A *uniformly gray* shark illuminated from above, as in natural lighting, would have a relatively bright back and a relatively dark belly. (B) A *countershaded* shark illuminated from the side, as in flash photography, stands out against the natural light behind it. (C) In a *countershaded* shark under natural lighting, the gradual transition from dark back to light belly has an averaging or cancelling effect. The top of the shark is seen as dark against dark, the middle as intermediate brightness against intermediate, and the light belly as light against light. All shark color-background combinations eliminate contrast between the shark and its background. Helfman et al. (2009); used with permission of Wiley-Blackwell

by their predators or prey from the side, and countershading works equally well then. When seen from slightly above the horizontal, the dark upper side of the shark absorbs relatively bright, mostly downwelling light, creating a dark target that is seen against the darker background of slightly upwelling light. Similarly, if viewed from slightly below the horizontal, a lighter-colored upper belly surface reflects weak upwelling light, creating a relatively bright target seen against the lighter background of slightly downwelling light.

Countershading makes even more sense when one realizes how conspicuous a uniformly colored shark would be. A uniformly light-colored shark viewed from above would reflect rather than absorb downwelling light and become a bright target against a dark background. Similarly, a uniformly dark shark viewed from below the horizontal would absorb weak upwelling light and appear as a dark object against a relatively bright background. Countershading is the best of all possible coloration worlds. With one minor exception. When viewed from directly below, any shark, regardless of color, is clearly visible as a dark silhouette against the background of bright, downwelling light.

Which sharks aren't countershaded?

Sharks that spend most of their time on or very near the bottom (wobbegong carpetsharks, scyliorhinid catsharks and shysharks, swell sharks)

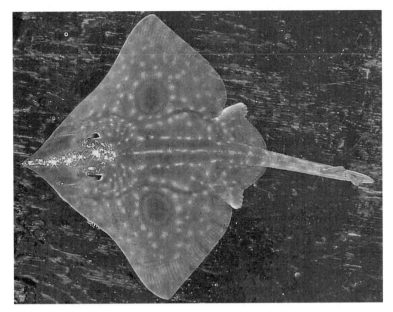

A Big Skate, the largest skate in North American waters. The round "bull's eyes" on the upper surface of the pectoral fins occur on many skates, but their function remains a mystery. Photo from Wikimedia Commons, www.afsc.noaa.gov/race/media/photo_gallery/fish_files/Big_skate.htm

may or may not be countershaded. Regardless, their backs, instead of being a solid dark color, are a mottled gray or brown, often arranged as lighter regions between dark saddles. In some sharks, such as triakid Leopard Sharks and scyliorhinid shysharks, the saddles and blotches wrap around almost the entire body before grading to a lighter belly. Such blotchy coloration is similar to that of benthic (bottom-living) bony fishes such as scorpionfishes, flatfishes, and sculpins. This color pattern most likely makes the animals blend into the mottled bottom against which they are seen from above.

Skates are generally countershaded, but their white bellies are almost always pressed against the bottom, unseen. In some skates, a pair of distinctive round bull's-eyes decorate the exposed top side; among these are the Eyespot Skate, *Atlantoraja cyclophora*; the Big Skate, *Raja binoculata*; the Starry Skate, *R. stellulata*; and the Longnose Skate, *R. rhina*. (See also the description of Epaulette Sharks in the next question.) How these "eyespots" function is a mystery. In many animals (toads, butterflyfishes, pufferfishes, Oscar cichlids, moths), eyespots on backs, fins, and wings are thought to startle or perhaps fool predators into thinking they are encountering a much larger animal. Alternatively, eyespots may invite an attack, deflecting it away from more vulnerable body areas, such as the head. Whether skate coloration serves a similar function is unknown.

Deviations from countershading in open-water sharks usually function to make a small part of the shark visible when first detected. These colored body parts probably separate one species from another, helping sharks tell

each other apart, a useful feature during mating. These seemingly minor patches of color are also helpful to people who are trying to tell sharks apart. This is the case with the Porbeagle (*Lamna nasus*). Porbeagles differ from all other lamnids because they have a distinctive white patch in the lower rear portion of their first dorsal fin.

These variations in coloration are indicated by the common names of many carcharhinid sharks: Blacktip Reef Shark, Blacknose Shark, Silvertip Shark (*Carcharhinus albimarginatus*), Whitetip Reef Shark, Whitenose Shark (*Nasolamia velox*). It appears that such color is meant to be obvious. The black tip on the dorsal fin of a Blacktip Reef Shark often has a white lower edge, which makes the black tip stand out even more. The Whitenose Shark's nose has a black tip with a white border behind the black, which makes it a blacknose shark, but that common name was already taken.

Tiger Sharks, with their noticeable vertical barring (a color pattern that fades with age) repeats a pattern seen in large bony fishes that often swim near the surface, such as King Mackerel, Wahoo, tuna, and marlins. The vertical barring is thought to mimic the flickering shafts of sunlight created by ripples at the water's surface, breaking up the outline of the fish and making it harder to identify. This type of camouflage might also explain the variably shaped white spots and lines on the backs of Whale Sharks.

Even more speculative are suggestions about the function of light color on the fin tips of Oceanic Whitetips. The ends of the first dorsal, pectorals, pelvics, and both lobes of the tail are distinctly white in this oceanic species, which feeds on fast-moving offshore fishes. The idea is that an otherwise countershaded shark approaching from a distance in open water is almost invisible, except for the white tips. The tips are made even more visible by side-to-side tail movements and adjustments of the pectorals. A potential victim could be fooled into thinking the obvious white marks were something small and maybe even edible, allowing the Oceanic Whitetip to get close enough to put on a burst of speed, when it was too late for the prey to escape. If you watch videos of these sharks, or at least videos that include footage as the shark approaches from out of the gloom, this idea makes sense.

Finally, deepwater sharks—those that live well below the depth of sunlight or moonlight penetration—tend to be uniformly dark in color. The Portuguese Dogfish, the deepest living shark, is uniformly blue-black or brown-black. This lack of patterning makes sense in terms of the assumed function of countershading and its link to light distribution nearer the surface: bright downwelling, intermediate spacelight, weak upwelling. In the inky-black depths, light is equally absent in all directions. This makes countershading, or any variation in color across the body, pointless.

Once again, known exceptions generally prove the rule. Some sharks

Sharks: The Animal Answer Guide

The Portuguese Dogfish is a deep-dwelling, 1-m-long (3-ft) somniosid shark with the uniform dark coloration typical of sharks that live in ocean depths beyond the reach of surface light. Photo courtesy of Maria Pickering

that live at intermediate depths and that move up and down day and night have dark backs while their lower surfaces are spotted with light organs called photophores. Spined Pygmy Sharks, diminutive dalatiids (28 cm, or 11 in, in length), occur around 500 m (1,640 ft) by day but move up to 200-m (660-ft) depths at night. They have numerous small photophores on their bellies right up to their snout. It is thought that these light organs put out just enough light of the correct wavelength to match weak downwelling light, thus obscuring their profile when viewed from below by their predators. This so-called "ventral bioluminescence" also characterizes many bony fishes, squids, and shrimplike animals that live at the same depths.

Are any sharks colorful?

Colorwise, sharks are pretty boring. Dominant colors are shades of gray and brown, some darker, some lighter. Colorful sharks that deviate from black, gray, and brown are definitely rare. Those that appear colorful—with spots, stripes, bars, or blotches—are almost entirely small bottom-living species that use their color patterns as camouflage, blending into the varied-color background on which they sit. Bold marks may break up the image of their body outline, making them unrecognizable as a shark. This kind of patterning, referred to as disruptive coloration, is common in many fishes, both predators and prey.

The squalomorphs, although taxonomically diverse, contain few sharks that could be considered colorful in any sense of the word. Some angel sharks (Squatinidae) have rings and spots on their bodies and pectoral fins. Angel sharks live on or even buried in sandy bottoms, and the spots probably serve (somehow) as camouflage.

Most galeomorphs are countershaded or unimaginatively colored,

but a few benthic (bottom-dwelling) families could be considered exceptions. Heterodontiform bullhead sharks, especially the juveniles, sport bold markings. Members of the genus *Heterodontus*, such as the Zebra Bullhead Shark (*Heterodontus zebra*—not to be confused with the even more striking stegostomatid Zebra Shark, *Stegostoma fasciatum*), have vertical bars, wavy lines, and numerous white or black spots. The skin coloration of orectolobid wobbegongs could serve as models for Realtree camouflage patterns. Wobbegongs sit motionless on algae-covered reef surfaces, and their color and head frills easily blend into those complex backgrounds. The hemiscylliid carpetsharks and bamboosharks are among the most colorful of all sharks, although their spots, blotches, and bars are still brown, black, or grey, though often set off with white edges. The tropical Pacific Epaulette Shark has a beige to brown back covered by brown spots and some darker bands. It gets its name from a very large black spot that is ringed by white on each "shoulder," behind the pectoral fins. (Epaulettes are shoulder bars on a soldier's uniform—and the red shoulder patches of Redwing Blackbirds.) The conspicuous ringed epaulette may draw a predator's attention away from vulnerable body parts, or it might be mistaken for a large eye that belongs to a much larger organism, one that shouldn't be messed with. We don't really know.

Also within the Galeomorphi is the strikingly marked Zebra Shark, juveniles of which are popular aquarium animals because of their pattern of brownish-black rings on a white background. These sharks outgrow even the largest home tank, becoming leopard-spotted adults 2.5 m (8 ft) long. Scyliorhinid catsharks can be split into two major groups as far as color is concerned. Many are uniformly colored deepwater animals (living at about 1,000 m, or 3,300 feet). The shallow-water catsharks are another matter altogether, with common names that reflect their coloration (Australian Marbled Catshark, Orange Spotted Catshark, Redspotted Catshark, Tiger Catshark, Pajama Shark). Many have multiple dark circles lined in white running the length of their body. Again, these are almost entirely bottom-sitters.

Other galeomorphs with variable coloration or shading are the Harlequin Catshark (*Ctenacis fehlmanni*), with orange-brown blotches and spots; the aptly named Leopard Shark of California kelpbeds; and the puzzling Atlantic Weasel Shark (*Paragaleus pectoralis*), which has a series of striking yellow bands running from its snout to its tail. The Atlantic Weasel Shark may be the only actively swimming shark that can be considered truly colorful, and its coloration sets it apart from all the bottom dwellers that are clearly camouflaged by their color patterns. Longitudinal (lengthwise) stripes, including yellow stripes, occur in a number of actively swimming bony fishes with relatively long bodies, such as snappers, rainbow runners,

goatfishes, and sweetlips, but how this color helps these fishes remains a matter of debate.

Perhaps strangest of all is the Goblin Shark (*Mitsukurina owstoni*), with its long, daggerlike snout, protrusible jaws, and slender, fanglike teeth. This deepwater (around 500 m) rare lamniform has a body that is soft pink to purple-gray, with blue-edged fins. Its color is due neither to pigments nor structural iridescence but to blood vessels that lie close beneath the skin, which is itself fairly transparent.

Rajid skates as a group are plainly colored, except for those with paired bull's eyes on their pectoral fins and backs. Many other skates, including the Spotback Skate, *Atlantoraja castelnaui*, have small black spots. Others, such as the Undulate Ray (*Raja undulata*) and the Sydney Skate (*Dipturus australis*), sport wavy lines and white spots reminiscent of Australian Aboriginal bark paintings, which may have been inspired by these batoids or by the color patterns on a Whale Shark.

Colorful stingrays are also the exception. Interestingly, several of these have blue spots and stripes. Blue is a pigment almost entirely absent from the selachian sharks, except for some open-water species like Blue Sharks and makos, whose countershading includes deep-blue backs. The Blue-spotted Stingray (*Dasyatis kuhlii*) of the tropical Pacific has blue spots on a greenish background. It is also one of the few rays thought to be able to see colors (see "Can sharks see color?" in chapter 2). The Bluespotted Ribbontail Ray (*Taeniura lymma*), also a tropical western Pacific reef species, has electric blue spots on a yellow background, with two blue stripes running the length of the tail. Some of the most colorful rays are the potamotrygonid freshwater stingrays of the Amazon Basin. The backs of these animals are spotted and streaked, often in (again) bright blue. Freshwater stingrays have powerful venom in their tail barbs, tempting an interpretation that their bright, contrasting colors serve to warn potential predators of these effective defenses.

One species of electric ray, the Ornate Sleeper Ray (*Electrolux addisoni*) of South Africa has a dark brown back with small yellow spots and curving, concentric black stripes. When it is approached by a diver, the ray arches its back, curls its disk, and raises its tail in what might be a threat display. Bold, obvious "warning" coloration and conspicuous body postures occur in a wide variety of animals with toxic properties, from wasps to caterpillars, poison dart frogs, and skunks. *Electrolux addisoni* also holds the distinction of probably being the only elasmobranch (or any animal?) whose genus name honors a brand of vacuum cleaner and reflects both its electrical traits and its suction-feeding behavior.

The large, graceful Spotted Eagle Ray has distinctive white, bluish-white, greenish, pearly, or yellow spots on a black background across its

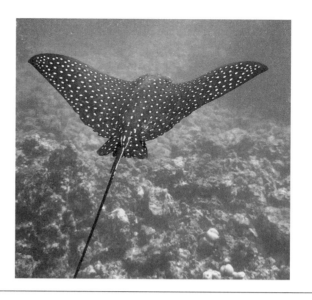

Spotted Eagle Rays are among the few rays that occur in tropical Pacific, Atlantic, and Indian Ocean waters. Often seen in small groups in relatively shallow water, these graceful, active rays are a favorite among divers. Photo by John Norton, Wikimedia Commons, http://en.wikipedia.org/wiki/File:Spotted_Eagle_Ray_(Aetobatus_narinari)2.jpg

back, including on its pelvic fins and the base of its tail. The spotting may continue onto the outer margins on the undersides of the wings. Individuals can apparently be told apart by their spot patterns. (Researchers also recognize manta ray individuals by the black spots on their otherwise white bellies.) Unlike most other stingrays, Spotted Eagle Rays spend little time resting on the bottom and much time "flying" around a reef.

Several chimaeras are relatively colorful, with deep reds, purples, and blues, colors seldom found in elasmobranchs. Their names reflect these color patterns: the Giant Purple Chimaera (*Chimaera lignaria*), the Purple Chimaera (*Hydrolagus purpurescens*), and the Pointy-nosed Blue Chimaera. How this coloration might interact with habitat and color vision remains a mystery.

Color usually comes from pigments or structures that reflect light of particular wavelengths (see "What causes the different colors of sharks?" below). But at least six different evolutionary lineages of deep-sea squaloid sharks have light organs on their sides that emit a greenish-yellow light, often in patterns specific to a species. Examples include the Velvet Belly Lantern Shark (*Etmopterus spinax*) and the dalatiid Cookiecutter sharks, the latter discussed in chapter 2 with respect to teeth and in chapter 7 on feeding habits. Cookiecutters can be considered micropredators or megaparasites, depending on your thinking. They live at middle ocean depths, where many squid, shrimp, and deep-sea fishes also have light organs. Tunas, dolphins, sperm whales, Megamouth Sharks, White Sharks, and other predators descend into this cloud of illuminated prey to feed. Cookiecutters are thought to hide among these other light producers, blending into the background of flickering bioluminescence, then sneaking up on the larger predators and removing plugs of flesh.

Sharks: The Animal Answer Guide

The Megamouth Shark, one of the victims of Cookiecutters, was at first considered a light emitter, producing a glow from its upper lip region that might attract plankton prey. This idea is unproven, either by observation of healthy live animals (unlikely) or by histological (microscopic) examination of mouth tissues, yet to be done. It is more likely that the distinctive white band of tissue along the upper jaw reflects dim light at midwater depths. The glow could attract prey animals, or it could be a species identification tag used by other Megamouths. In truth, we don't know the function of the unique white oral band in this species.

Are any sharks colorless?

Albinism occurs in animals because of a malfunction in the gene responsible for producing melanin pigment. A true albino shark is almost white (in underwater photos albinos appear to glow) and has pink eyes. Such specimens are rare and most often involve embryos or newborns. Few true albinos live long because they lack the coloration necessary to protect them from a variety of large, visually hunting predators. The rarity also means that just about anytime someone finds an albino shark, or even a partial albino, the discovery gets published in a scientific journal.

Albino embryos (often removed from a dead shark's uterus) or very young albino individuals are known in the Gray Smoothhound Shark (*Mustelus californicus*), the Whitespotted Bamboo Shark, the Leopard Shark, the Sandbar Shark, the Tiger Shark, the White-Spotted Marble Electric Ray (*Torpedo marmorata*), and the Common Torpedo (*Torpedo torpedo*). Albino Brownbanded Bamboo Catsharks have been born to a captive, normally colored female at the Denver Downtown Aquarium. The fish collection at Rhodes University in Grahamstown, South Africa, has a real white shark, or rather a very young albino White Shark 1.7 m (5.6 ft) long. Most bizarre is a 22-inch-long (56-cm-long) albino Dusky Shark (*Carcharhinus obscurus*) embryo that was removed from a pregnant female by a fisherman in the Sea of Cortez, Mexico, in 2011. What made this find truly strange was that it was a cyclops. It lacked normal eyes but had one fully developed, forward pointing eye between its mouth and snout.

Full-grown albino sharks do occur and are even sometimes photographed in the wild. Species include a Narrownose Smooth-hound (*Mustelus schmitti*), an adult Tawny Nurse Shark (*Nebrius ferrugineus*), and a Halave's Guitarfish (*Rhinobatos halavi*). An albinistic adult male Sand Tiger was filmed off the New South Wales coast of Australia in 2007, and the same month, a 10-m (33-ft) female albino Whale Shark was glimpsed and even videotaped near Darwin Island, Galapagos. That same year—a seemingly big year for albinos—a 12-inch-long (30-cm) juvenile female albino Spot-

Yes, Virginia, there really is a *white* shark. This 1.7-m-long (5.6-ft) albino White Shark was caught off Boknes in Eastern Cape Province, South Africa, in 1996. It is the only known albino White Shark and in fact the only known albino member of the family Lamnidae. Normally colored White Sharks get their name from their very white bellies. Photo by Gene Helfman

ted Ratfish was caught not far from Seattle, Washington. An adult male albino skate (mislabeled as a stingray in the posting) was captured by a fisherman in British waters in 2011. Albino Cownose Rays (*Rhinoptera bonasus*) show up often enough that SeaWorld Orlando has a live one on display. A few albino Short-Tail Stingrays (*Dasyatis brevicaudata*), perhaps the world's largest stingray, have been seen and photographed. Some other photos appearing on the web include an albino Zebra Shark (size not given) photographed at East Wallabi Island, Western Australia. A presumed albino juvenile hammerhead shark in the Cebu Zoo, Philippines, turned out to be a pangasiid catfish.

What causes the different colors of sharks?

Sharks have skin coloration of two types. Most common are skin colors due to pigments in the skin cells. Melanin pigment results in grades of black; guanine results in grades of white. The color-containing cells are called chromatophores (literally, "color bodies"), defined by the pigments they contain. For example, melanin is concentrated in color cells in the skin called melanophores. Color change occurs when the shark expands or contracts its chromatophores. When melanophores contract, those regions of the shark become lighter in color. Pigments are either produced by the shark itself or somehow transferred from colorful food organisms.

Less common are structural colors that cause iridescent (shiny, glowing) color. Iridescence results from light passing through and reflecting from

Sharks: The Animal Answer Guide

thin surface tissue. Iridescent colors change depending on the direction of incoming light and the angle at which a shark is viewed.

The Smalleye Hammerhead (*Sphyrna tudes*) of eastern South America is sometimes called the Golden Hammerhead. Juveniles sport both structural and pigment coloration that makes them gold. Orange and yellow pigments in the skin cells add brightness to the color, which is also enhanced by iridescent structural tissues. The result is that live animals have glowing, metallic golden heads, sides, and fins that lose their glow upon death. Even more interesting, the yellow and gold pigments are not produced by the animal itself but apparently come from its diet, in particular, a kind of shrimp and a marine catfish and its eggs. The Yellow Smoothhound (*Mustelus higmani*) feeds on the same shrimp and is also yellowish. Horn Sharks in California are another example of shark "coloration" obtained from an external source. Horn sharks feed heavily on purple sea urchins and often have purplish-pink teeth and fin spines.

After death, the drying and shriveling of the skin alters the thin surface tissues. Hence, iridescent colors quickly disappear in dead sharks. Pigment-based colors fade more slowly as the pigment granules decay with time. The fairly uniform yellowish-brown color (with some darker and lighter areas) of sharks that have been dead for a while, such as those in museum displays, doesn't do justice to their colors in life.

One interesting twist on color change occurs in deepwater sharks like the Velvet Belly Lantern Shark. Lantern sharks are uniformly dark but have small light organs (photophores) on their bellies (see "Which sharks aren't countershaded?" above). Lantern Sharks change the amount of light they emit in much the same way that shallow-water sharks change their color by controlling melanin distribution—except that Lantern Sharks use melanin to control light emission. On top of each photophore is a closeable diaphragm filled with melanin that works like the iris diaphragm in our eyes. The light stays on all the time, but the Lantern Shark contracts and expands the melanin in the overlying diaphragm to allow light to shine out or not.

Is there a reason for the color patterns of sharks?

Most fishes, either predator or prey, are colored in a way that makes them difficult to see underwater. "Invisibility" is a good explanation for the dominant color patterns of sharks, whether countershaded or bottom-matching. Countershaded sharks blend into the background of the water column, and mottled-back sharks and rays are camouflaged against the background of the ocean floor. In benthic sharks, skin growths and folds increase the resemblance to the bottom, as in the orectolobid wobbegongs.

If sharks are so well camouflaged by countershading, you can reasonably ask why photographs of sharks and rays often make them seem quite visible. The answer is that photos are usually taken with artificial lighting distributed in unnatural ways, at least as far as the evolution of shark coloration is concerned. Underwater photographers usually use strobe lights, which shine or flash very bright light on the side of a shark and thus illuminate it unnaturally, and conspicuously. The distribution of light on a dock or boat deck also misrepresents the real visual world of sharks. We live in a terrestrial world where light is more evenly distributed above and to the sides, and is often brighter from the side than from above. Underwater light, by comparison, is predominantly (95%) downwelling, with only about 4% of light coming from the side. Countershaded sharks are colored to match this light distribution and thus become invisible (see "Why are so many sharks dark on top and light on the bottom?" above). Add to this the fact that photographers seldom take pictures of well-camouflaged sharks, the ones that you can't see. Or at least those photos don't make it into books.

What color are a shark's eyes?

Eye color in sharks and all vertebrates is primarily a function of structural colors in the iris, the area surrounding the pupil. The iris of many sharks (Tiger, White, Porbeagle, makos) is large and black, or whitish (many carcharhinids), although a few species (Milk Shark, *Rhizoprionodon acutus*) have a yellow iris, and others (Leopard Shark) have a lighter, gray iris around a black pupil.

The pupil itself in most sharks is black and round (sixgill sharks, Blue Shark, White Shark, hammerheads, Sand Tiger), although some sharks have a vertical, slit-like pupil (Lemon Shark, Blacktip Reef Shark, Silky Shark), and in others the slit is horizontal or nearly so (some heterodontid bullhead sharks, squatinid angel sharks, Leopard Shark, Small-Spotted Catshark). Exceptions to black pupils are found in several deep-sea sharks that have large green pupils, such as centrophorid gulper sharks, etmopterid lanternsharks, and somniosid dogfish and sleeper sharks.

The "eye color" of some batoids is confusing because the pupil may be covered by a variously pigmented curtain called the operculum pupillare (literally "pupil cover"), an eyelid-like finger of tissue that expands over the eye in bright light conditions. Examples include the Banded Guitarfish (*Zapteryx exasperata*), many skates, sawfishes (Pristidae), and dasyatid stingrays. The highly reflective eyes of holocephalan chimaeras generally have a dark pupil surrounded by a lighter iris.

Do a shark's colors change as it grows?

Some sharks change color as they grow, with younger sharks often but not always more colorful than adults. Color change can be subtle and may involve loss of bars or loss of color in the tips of fins. For example, the vertical barring that gives Tiger Sharks their name is much more obvious in younger animals. The Zebra Bullhead Shark has zebra stripes at all stages, but the stripes are reddish-brown in juveniles and become black to dark brown in adults. Among the more impressive color changers are the hemi-scyliid bamboo and carpetsharks, which are banded and spotted as juveniles but more uniform brown or only slightly banded as adults. For example, juvenile Epaulette Sharks have light and dark bands over their body and fins; these bands break up and become more spotlike later. All stages have the distinctive white-ringed epaulette on the shoulders. The most dramatic color changer may be the Zebra Shark, which changes enough with age that juveniles and adults were once thought to represent different species. Juveniles have distinctive light or yellow zebra stripes or bars against a dark background, whereas adults have dark spots on a light background.

An exception to the trend of decreased coloration is the Leopard Shark of California kelp beds, which is slightly more colorful as an adult. Adults have black saddles and large black spots along the back; juveniles have fewer spots and saddles.

Nurse Sharks darken as they grow. There is some evidence that female

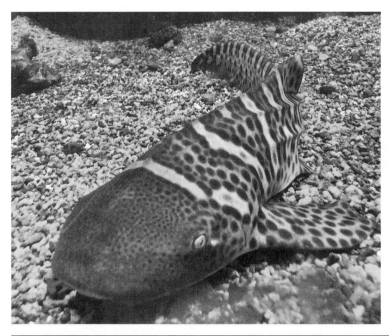

Juvenile Zebra Sharks have a banded color pattern; adults are covered in spots. This fish is a large juvenile that is changing from one to the other and is intermediate in color, showing both bands and spots. From Wikimedia Commons, http://en.wikipedia .org/wiki/File:Stegostoma_fasciatum_01_by _Line1.JPG

Nurse Sharks pay attention to color differences when choosing mates, at least to the extent that they avoid lighter-colored males during mating activities. The role played by absolute male size in these observations may complicate the issue.

Do a shark's colors change at different times of the day or in different seasons?

Bony fishes, especially tropical marine and freshwater species, change color on a moment's notice, or exhibit very different colors during the day than at night, being mostly blotchy at night regardless of their daytime hues. Background matching is well known in bony fishes, a classic example being a flounder's ability to almost mimic a checkerboard when placed on that kind of surface.

Most sharks, however, remain the same basic color throughout the day, at night, and in different seasons. A few can change color slowly, perhaps to better match their background, becoming darker or lighter or more blotchy. Obviously, this is more useful among benthic sharks than in species that swim actively in the water. Parascylliid Collared Carpetsharks (*Parascyllium collare*) are reported to change their skin color to better match seafloor coloration. Juvenile Nurse Sharks held in tanks lightened in skin color over a period of minutes when the tank they were kept in was covered and darkened. Little Skates held in illuminated tanks with white or black walls also changed color, becoming light brown in white tanks and dark brown in dark tanks. The color change took on the order of 9 to 12 hours, much longer than in color-changing flounders and other bony fishes. Spiny Dogfish also get lighter when kept in white-walled tanks, but the color change takes even longer, about two days. Physiological studies in skates and bottom-living sharks such as triakid smoothhounds and Small-spotted Catsharks showed that the animals sensed outside light using their eyes and the so-called third eye or pineal organ on the top of their heads. Skin color changes were due to a melanophore-stimulating hormone secreted by the pituitary gland that caused melanophores in the skin to expand or contract.

Is there much geographic variation in the color of a shark species?

Geographic variation in color occurs in sharks, although many species that were thought to be geographic variants turn out to be different species. Our increase in understanding comes from additional study, including improved genetic techniques that reveal differences that weren't obvious from external anatomy alone. For example, the Puffadder Shyshark of

Sharks: The Animal Answer Guide

South Africa was thought to come in two forms, Cape and Natal, that differed mainly in the color of their dark saddles and white spotting. A study of mitochondrial DNA genes showed that these are in fact two species, *H. edwardsii* and the Natal Shyshark, *H. kistnasamyi*.

Just the opposite applies to another scyliorhinid, the Saddled Swellshark (*Cephaloscyllium variegatum*) of the east coast of Australia. When the discoverers originally described this shark, they thought it was part of a three-species complex, with the species differing in their saddles and spots. But the coloration followed a latitude trend from north to south, with darker saddles in tropical individuals than in temperate animals. *C. variegatum* is now recognized as one species. Its species name, *variegatum*, refers to this within-species color variation.

Porbeagles are high-latitude lamnids that seldom enter tropical waters. Their coloration differs in northern and southern areas, evidence that little genetic exchange occurs between populations. Porbeagles in the Northern Hemisphere are mostly white under their head and abdomen, whereas the undersides of the heads of Southern Hemisphere adults may be dark, with dark blotches on an otherwise white abdomen.

Chapter 4

Shark Behavior

Are sharks social?

As a rule, sharks, skates, rays, and chimaeras tend to be solitary animals. When they come together outside of the breeding season (on reproduction, see chapter 6), it's most likely a result of the mutual attraction of individuals to food. So, for example, White Sharks will gather to feed on dead whales; Gray Reef and Silky sharks will home in on injured fish; and Basking and Whale sharks and Manta Rays will converge on dense patches of zooplankton. Sometimes, as happens in the Maldives during the southwest monsoonal season, hundreds of Reef Manta Rays and dozens of Whale Sharks aggregate at the same areas to feed on the same dense swarms of krill (small crustaceans). Although uncommon, some aggregations of sharks can result in a classic feeding frenzy. Feeding frenzies are antisocial gatherings, given that everything in the area can be bitten, including other sharks.

However, some species form small, apparently social groups, often when they are not feeding. Species that rest by day and feed by night may aggregate during their daytime resting period and then disperse to feed at night. Adult Small-spotted Catsharks gather in all-female groups in shallow, inshore caves during the day, in large part to avoid being harassed by male sharks. Small groups of Port Jackson Sharks (*Heterodontus portusjacksoni*) rest in caves during the daytime, disperse to feed at night, and then return to the same cave the next morning. Nurse Sharks spend the day in inactive groups of up to 40 individuals, also dispersing at night to feed and returning each morning to the same resting site. Some Whitetip Reef Sharks follow a similar pattern of daytime, grouped resting in caves and

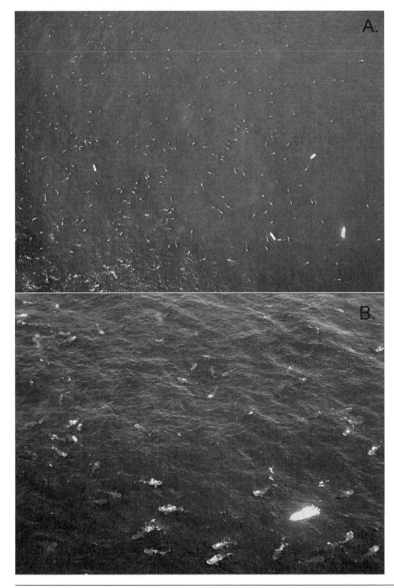

The world's largest aggregation of Whale Sharks, the biggest fish in the sea. On August 12, 2009, 420 Whale Sharks were spotted off the Yucatan Peninsula of Mexico in an area about 3 km × 6 km (1.9 × 3.8 miles). The sharks were feeding on recently spawned eggs of Little Tuna. The upper photo (*A*) shows 220 Whale Sharks and four tour boats; the lower photo (*B*), from a lower altitude, shows 68 Whale Sharks, one boat, and two pairs of snorkeling tourists.

From Rafael de la Parra Venegas et al., "An unprecedented aggregation of Whale Sharks, *Rhincodon typus*, in Mexican coastal waters of the Caribbean Sea," *PLoS ONE* 6, no. 4 (2011): e18994; photo used by permission of Alistair Dove / Georgia Aquarium

nighttime dispersal, although groups of Whitetips have been filmed feeding at night.

Aggregation in sharks doesn't necessarily mean lying on the bottom in the manner of Port Jackson, Nurse, and Whitetip Reef sharks. Actively swimming sharks form daytime resting aggregations too. Aggregations of Blacktip Reef Sharks are frequently observed milling around in water so shallow that their dorsal fins stick out; what they're doing is not really known, but their behavior does not appear to involve active feeding. Young Blacktip Sharks (*Carcharhinus limbatus*) also aggregate by day and disperse at night, again returning to the same daytime locales over a period of three years.

Scalloped Hammerheads swim slowly above seamounts (small peaks in the ocean's floor) by day, in groups as large as 500 sharks. They disperse as solitary individuals at night to feed, sometimes as far as 20 km (12 miles) away and then return the next morning to the same seamount. The best-known Scalloped Hammerhead aggregation is one that has been studied at Espiritu Santo Seamount, off the coast of Baja California, Mexico.

Gray Reef Sharks also form daytime aggregations in Pacific atoll lagoons. Dozens of animals may mill around in a small area inside the lagoon or swim in place against currents in a lagoon pass during the day, stacked up like logs along the walls of the pass. At dusk, they leave via the pass to forage outside the reef, returning again the next dawn to the same daytime resting locale. Daytime resting aggregations followed by nighttime dispersal are also common among bony fishes that feed at night on coral reefs and in lakes.

Rays and skates are also mostly solitary animals. Exceptions include some rays that rest in small groups on the bottom, such as Cowtail Stingrays (*Pastinachus sephen*) and Reticulate Whiprays (*Himantura uarnak*), as well as myliobatid rays (Bat, Cownose, Spotted Eagle, and manta) that swim up in the water column in groups of varying sizes. To the best of our knowledge, which is admittedly limited, chimaeras are basically solitary animals.

Do sharks form schools?

A school is a coordinated group of animals moving in the same direction at the same speed, often separated by a predictable but small distance. By this definition, most sharks do not school, but a few definitely do. In such schools, most individuals are of a similar size and the same sex, except of course during mating, when many otherwise solitary sharks form large groups of both sexes. Mating aggregations are discussed in chapter 6.

Among the best studied of the schooling sharks are Scalloped Hammerheads. Males tend to aggregate inshore, whereas females congregate offshore over deeper water, sometimes in groups of several hundred animals. Female schools tend to mill around seamounts, rather than moving continually in one direction. School Sharks, or Tope (*Galeorhinus galeus*), often move in small schools composed mostly of individuals of the same sex and age. Blue Sharks swim in groups segregated by sex even before the animals mature, with females staying nearer coastlines than males. Blue Shark groups are usually small but sometimes occur in large numbers; as many as 1,000 Blue Sharks have been seen near the surface at the edge of the continental shelf off the coast of Maryland. By the strict definition of schooling, these are more likely aggregations rather than true schools. Other species

that are frequently filmed in groups include Blacktip, Blacktip Reef, Silky, Porbeagle, Whitetip Reef, Gray Reef, Sand Tiger, and Basking sharks.

Spiny Dogfish—probably the most abundant shark in the sea—form schools numbering in the hundreds and even thousands. These schools create problems for fishing operations because migrating schools move through areas and destroy fishing gear, get tangled in nets set for other fishes, eat large numbers of fish and invertebrates, and may slow the recovery of depleted stocks of fishes such as Atlantic Cod. Spiny Dogfish schools are usually made up of animals of similar size and, after the fish mature, of one sex. Immature dogfish tend to school offshore whereas mature females school farther inshore.

Among batoids, the myliobatid rays are well known for forming groups. Manta Rays and Spotted Eagle Rays frequently occur in small groups of a half dozen or more animals, swimming in synchrony. Spotted Eagle Rays fly through the water in small groups of 5 to 50 individuals. The groups take on predictable shapes, including one ray following another or the group swimming in a diamond formation, much like planes at an air show. Manta Rays in Hanifaru Bay, the Maldives, can form trains of dozens of animals that perform what has been called cyclone feeding. The mantas form a giant linked chain of more than 50 animals, the ray at the end of the chain followed by the ray at the head of the chain, forming a spiraling vortex of Manta Rays.

The record holders for large schools of sharks are smaller myliobatids, such as the Bat Ray (*Myliobatis californica*) and especially the Cownose Ray. Migrating hordes of Cownose Rays, called "fevers" when migrating in large groups, have been counted from airplanes. One fever in Sarasota Bay, Florida, was estimated to contain between 600,000 and 1 million individuals; another group, in the Chesapeake Bay of Virginia, was so large that numbers had to be calculated from multiple aerial photographs. This migration contained more than 5 million individuals, covering an area of over 450 ha (1,110 acres, or more than 850 football fields), at a density of more than 1.1 ray per square meter (or yard).

Can sharks tell one another apart?

Most sharks and rays clearly can distinguish between their own and other species. This is shown in the many species that form aggregations and schools of only one species, often of only one sex, and mostly made up of individuals of similar sizes. In laboratory experiments, juvenile Lemon Sharks preferred to associate with other Lemon Sharks rather than with Nurse Sharks, and Nurse Sharks preferred to be with their own species rather than with Lemon Sharks.

But within these groups, can sharks identify different individuals? Behavioral studies have shown that certain sharks and rays occur or reappear at the same place for months or even years, often together with the same individuals. Examples include White Sharks at the Farallon Islands near San Francisco, at Guadalupe Island in Mexico, and near Cape Town, South Africa; Small-spotted Catsharks and Port Jackson Sharks in daytime resting caves; Sicklefin Lemon Sharks in Tahiti; Blacktip Reef Sharks in Moorea, French Polynesia; Whitetip Reef Sharks in Hawaii; Zebra Sharks in Queensland, Australia; Whale Sharks in the Maldives; Spotted Eagle Rays at Bimini in the Bahamas; and Reef Manta Rays in Mozambique.

Researchers identify individuals using natural marks on fins, colored spots and scars on the body, fin tears and notches, or distinct color patterns. For example, in Blacktip Reef Sharks the relative size and shapes of the black, brown, and white patches at the tip of the dorsal fin differ among individuals. Individual Zebra Sharks have unique color patterns, and Spotted Eagle Ray individuals have unique spot patterns. Photos over many months show that the differences are long-lasting, identifiable for up to a 10-year period in Blacktip Reef Sharks. In some species such as White Sharks, the accuracy of the identifying marks has been confirmed with genetic fingerprints. This allows researchers to accumulate photo albums of individuals and follow them through time, as has been done for humpback and killer whales.

Do the sharks also tell one another apart? This is a harder question to answer than simply asking if the same animals occur together repeatedly. Some evidence suggests that sharks are capable of selecting the sharks they associate with, at least when kept in captivity in small numbers. Female Small-spotted Catsharks kept in small groups in laboratory tanks tended to associate with particular individuals and avoided others. Their behavior changed notably when unfamiliar females were introduced into a group, again suggesting that they knew who was who. Whether similar behavior in this species occurs in the wild is unknown. Observations of 38 wild juvenile Lemon Sharks at a known aggregation site at Bimini, Bahamas, indicated that certain individuals tended to associate preferentially with other individuals. Pairs tended to be made up of similar-sized animals, regardless of sex, which could be individual recognition or just a preference for associating with an animal of the same size. Researchers watched from nearby towers and could tell individuals apart because the sharks had been tagged with color-coded tags.

More definitively, a careful study of a population of 133 Blacktip Reef Sharks in Moorea demonstrated that the population was divided into four subgroups of 14 to 54 individuals per group. Individuals within the groups tended to pair with some individuals and avoid others, with animals of a

Using variation in fin coloration in Blacktip Reef Sharks to identify individuals. (*Top panels*) Left and right sides of a single female (*A*) and male (*B*) shark. (*C–H*) Photos from left and right sides of the dorsal fins from six other known sharks. (*I–J*) Photos of the same two individuals taken 10 years apart, showing that color patterns remain constant.

A–H, photos by Johann Mourier; I–J, photos by Ila Porcher and Johann Mourier; published in J. Mourier, J. Vercelloni, and S. Planes, "Evidence of social communities in a spatially structured network of a free-ranging shark species," *Animal Behaviour* 83 (2012): 389e401; used with permission

similar size and the same sex associating strongly, resulting in "long-term preferred and avoided companions." The strength of these associations was comparable to the complex social structure of such highly social marine mammals as dolphins and sea lions. Whether the sharks use the same visual cues that we use to tell each other apart or some other cues remains to be determined.

Do sharks fight?

Our knowledge of behavioral interactions in sharks has grown in recent years with increased time and effort by researchers. We can generalize that when two sharks meet around food, larger animals will generally dominate smaller animals. Direct contact and fighting is rare; this makes sense, given the damage that sharks can inflict on one another. Often, an interaction ends when a smaller animal simply moves away as a larger animal approaches. Such size-determined outcomes have been observed in Smooth Dogfish; Bull, Sandbar, Lemon, and Galapagos sharks (*Carcharhinus galapagensis*); Bonnetheads; and White Sharks. In Bonnetheads, sex as well as size determines behavioral interactions, with larger sharks and males dominating smaller sharks and females. Juvenile Largetooth Sawfish (*Pristis microdon*) in captivity use their toothed rostrum ("saw") in battles over food.

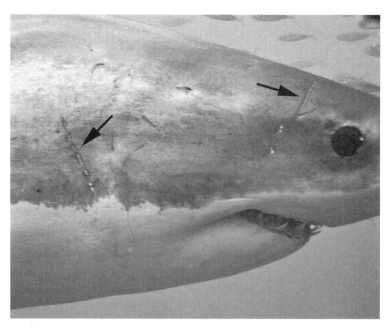

An approximately 3.5-m-long immature White Shark swims past a shark cage at Dyer Island, South Africa. This shark has apparent puncture-like bite marks on its head (*arrow at left*) that were most likely inflicted by another shark. It also has elongate scars of the same shape and dimension as the wire mesh from which the cages are constructed (*arrow at right*), suggesting that the shark forcefully stuck its head inside a cage at some time in the past. Photo by Gene Helfman

A larger sawfish lays its rostrum across the rostrum of a smaller sawfish, pinning it to the bottom and forcing it to give up a meal.

In White Sharks, size definitely matters, and small sharks defer to larger animals. But when two sharks of about the same size interact around food, they engage in several behavioral displays. Researchers have catalogued these behaviors while watching sharks around floating baits near Dyer Island, South Africa. One shark may follow another with its mouth open, and the other will swim away; or two sharks will swim at one another until one veers off in a White Shark version of "chicken." Sometimes a shark will roll on its side and slap its tail at another nearby animal, which may return the tail slaps. Such "ritualized" fighting without actual physical contact is common throughout the animal kingdom and is considered a means of settling disputes without injury. If neither animal retreats, biting may occur, and some White Sharks bear obvious bite marks from such encounters. Non-mating injuries, including missing fin parts, are not uncommon in many species of sharks. It is likely such interactions within species are much more common than has been formally documented.

Although battles among stingrays are seldom observed, we have evidence that they will fight over food. An area in the Cayman Islands is known as Stingray City because of the large numbers of Southern Stingrays that gather there to be fed by divers. This is a normally solitary species, unaccustomed to associating with many members of its own species. Handouts are not always available, and the rays jostle for position and

Sharks: The Animal Answer Guide

food, occasionally biting one another. Rays at Stingray City have more bite marks on their bodies than rays that live in nearby areas where they aren't fed by tourists (see "Should people feed sharks?" in chapter 8).

Are any sharks territorial?

Territoriality usually means defending a locale against intruders; a damselfish may defend its breeding and feeding territory against other fishes, or a dog may defend its yard against other dogs or people. In this sense, we have little evidence of territoriality in sharks. Some sharks react when other sharks or people come too close, a behavior that could be called defense of personal space. This defense was first described in Gray Reef Sharks.

Gray Reef Sharks have a distinctive, predictable reaction to being approached too closely, whether or not they are feeding. The approached shark lifts its snout, arches its back, lowers its pectoral fins, and swims in an exaggerated manner, swinging its head back and forth. If the intruder doesn't back off, the displaying shark may then attack. An unfortunate diver was first to describe the sequence, after being bitten by a Gray while trying to get close enough to take a good photo. This then led researchers to work out the details by deliberately approaching sharks in very small submarines. Sharks attacked the submarines also.

Something between such defense of personal space and possible territoriality has been observed in the large resting aggregations of Scalloped

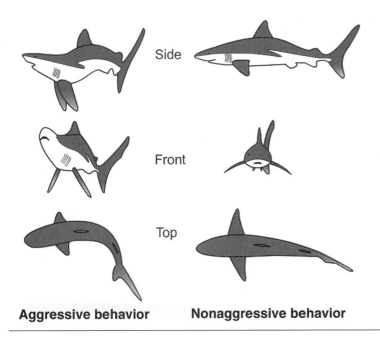

Aggressive behavior **Nonaggressive behavior**

Exaggerated swimming display of the Gray Reef Shark when its personal space is invaded by a diver, another shark, or a small submarine. Exaggerated body and fin positions of an aggressive shark are shown on left, with normal swimming postures on the right. Redrawn by Chris Huh, Wikimedia Commons, http://en.wikipedia.org/wiki/File:Shark_threat_display.png, based on R. H. Johnson and D. R. Nelson, Agonistic display in the Gray Reef Shark, *Carcharhinus menisorrah*, and its relationship to attacks on man," *Copeia* 1973, 76-84; used with permission of the American Society of Ichthyologists and Herpetologists

Hammerheads mentioned above. Among Scalloped Hammerhead females, central positions within the group seem to be favored, and these positions are most often occupied by larger animals. Spacing between sharks and position in the group are enforced by several displays. Larger females displace smaller animals by swimming toward the smaller shark and performing an underwater spinning backflip, referred to as "corkscrewing," something worth trying at home. Smaller females respond to this display by moving away, usually towards an outer position within the group. At a higher level of aggression, a large female may swim over a smaller one and strike down with her head on the head of the smaller shark in a motion similar to the way that hammerheads "hammer" stingrays into the bottom (see "How do sharks catch their prey?" in chapter 7). Again, the smaller individual usually moves off to the edge of the group after being struck.

Displays similar to those of the Gray Reef Sharks have been seen in Galapagos, Silky, Lemon, Blacknose, and Bonnethead sharks; and a Bull Shark performed exaggerated swimming before attacking a small boat that had been following it. These sharks and Scalloped Hammerheads are all members of two closely related families, the Carcharhinidae and Sphyrnidae, so it would not be surprising if other species in these families also maintain personal space via similar displays. Defense of personal space may also explain a behavior known from Sand Tiger Sharks. Female Sand Tiger Sharks off the Natal Coast of South Africa perform a tail whipping behavior that sounds like a whip cracking. This tail whip-cracking has been directed at divers who approached the sharks in caves and could be a keep-away signal.

Basically, if you approach a shark and it starts to act strangely, consider it a warning you shouldn't ignore.

If we expand the definition of a territory to include defending an object rather than just a piece of turf, we can include defense of food as a type of territoriality. Sharks will defend food against other sharks. White Sharks perform displays toward other White Sharks around food, including parallel swimming and slapping the tail on the surface in the direction of another, approaching White Shark.

Home ranges are often distinguished from territories. A home range, sometimes called an activity space, is the area where an animal spends most of its time, regardless of whether it defends it or not. Many shark species have well-defined if not constant home ranges. Horn Sharks constrain their activities to a relatively small home range of about 1,000 m² (11,000 ft²), which is about twice the area of a basketball court. The area includes a shelter site for daytime resting and may be used by an individual for a decade or more. Some other sharks with known home ranges include Blacktip Reef Sharks (about 0.5 km², or about 0.2 square miles) and Gray Reef

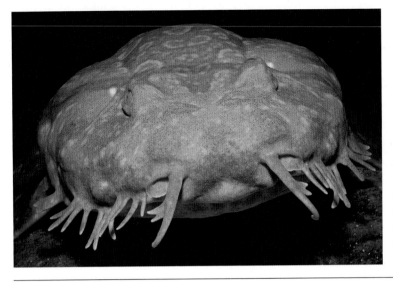

A Wobbegong, or carpet shark. The skin flaps around the mouth and variable body colors add to its camouflage as it lies on the bottom. Photo courtesy of Jeff Rotman, www.jeffrotman.com

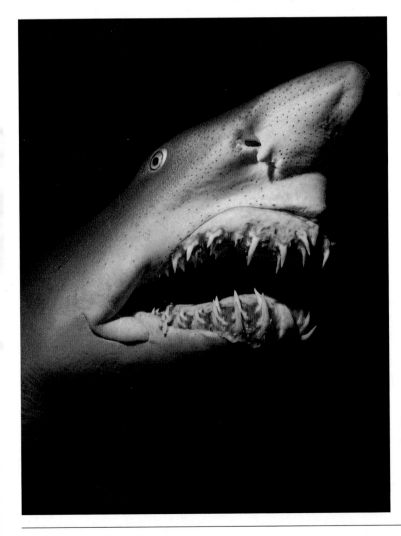

Sand Tigers feed on slippery prey such as fishes and squid. Teeth in both jaws are long, slender, and sharp, the ideal type for stabbing and holding such food items. Photo courtesy of Jeff Rotman, www.jeffrotman.com

A Puffadder Shyshark, *Haploblepharus edwardsii*, at the Two Oceans Aquarium in Cape Town, South Africa. The mottled color pattern of the body would make this shark blend well into the natural dark, algae-covered bottom on which it is normally found. The object above the shark between the second dorsal fin and tail is its egg case, or mermaid's purse. Photo by Gene Helfman

A Tiger Shark moves across shallow sand flats in the Bahamas. Notice the faint vertical bars on the shark's back, markings that may help camouflage it in shallow water and flickering sunlight. These markings fade as Tiger Sharks grow. The shark is being followed closely by a remora, or sharksucker, swimming beneath it. Photo by Albert Kok, Wikimedia Commons, http://en.wikipedia.org/wiki/File:Tigershark3.jpg

An Oceanic Whitetip Shark accompanied by a Pilot Fish. The white tips of the shark's fins are thought to help conceal the rest of the body as the shark approaches prey. Whitetips are open-water sharks that seldom come into shallow water, fortunately. They have been implicated in a number of attacks on victims of offshore airplane crashes and boat sinkings. Photo by Austin Gallagher, http://austingallagher.com; used with permission

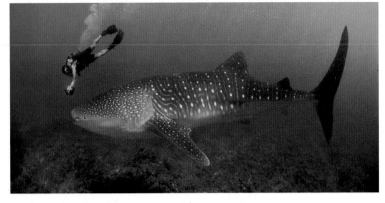

Whale Sharks are noted for their distinctive white spots and wavy lines on a dark background, color patterns eerily similar to the bark paintings of Aboriginal Australians. This 7-m-plus (22-ft) Whale Shark was being filmed at Ningaloo Reef, West Australia, the first place Whale Sharks were studied in any detail. Photo courtesy of Jeff Rotman, www.jeffrotman.com

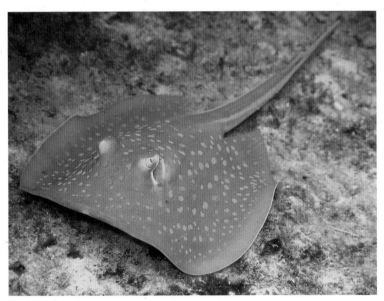

The Bluespotted Ribbontail Ray is a relatively small (35 cm, or 14 in, wide) inhabitant of shallow waters in the tropical western Pacific. Its small size and color make it popular in the aquarium trade, but individuals seldom live for long in captivity. Photo by Jens Petersen, Wikimedia Commons, http://en.wikipedia.org/wiki/File:Taeniura _lymma2a.jpg

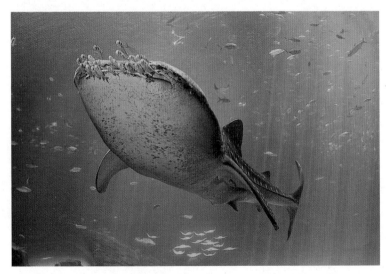

Golden Trevally swim just ahead of a Whale Shark at the Georgia Aquarium in Atlanta. This same association has been observed in the wild. Benefits to the trevally could include improved food finding or energy saving; benefits to the shark are probably minimal. Photo by Marj Awai, courtesy of the Georgia Aquarium; used with permission

An aggregation of Scalloped Hammerheads. This species congregates over seamounts by day in groups that can number in the hundreds. Individuals then move away from the seamount to forage at night.
Photo by Seawatch.org, Wikimedia Commons, http://en.wikipedia.org/wiki/File:Sphyrna_lewini_school.jpg

A Common Thresher Shark, one of three species of thresher sharks. Half of the length of a thresher is taken up by the tail, which the sharks use to whack the small fishes on which they feed. Photo by Petter Lindgren, Wikimedia Commons, http://en.wikipedia.org/wiki/File:Threshershark.jpg

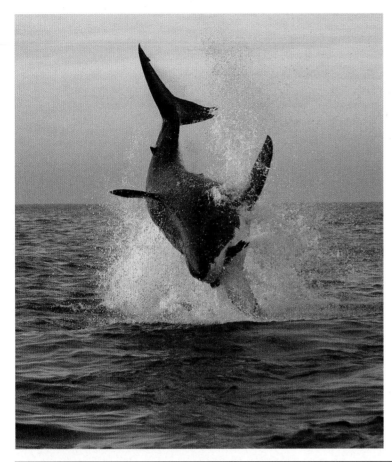

"Polaris breach!" A large White Shark cartwheels through the air after launching itself upward to feed on a Cape fur seal. In this case, the shark was fooled by a decoy shaped like a fur seal that was being towed behind the photographer's boat. The shark is still holding the decoy in its mouth. Photo by Austin Gallagher, http://austingallagher.com; used with permission

Sand Tiger Sharks are often kept in large public aquariums because of their ability to survive captivity. This more than 2-m-long (6.5-ft) male was one of several Sand Tigers in the Two Oceans Aquarium, Cape Town, South Africa. Photo by Gene Helfman

Basking Sharks got their name because they often swim slowly near the surface while feeding, behavior that was mistaken for basking in the sun. They also do not avoid approaching boats, making them easy to spear or ram. This harassment and serious overfishing have caused their populations to crash world-**wide.** Photo courtesy of Jeff Rotman, www.jeffrotman.com

(*Top*) The Spotted Ratfish, *Hydrolagus colliei*, an abundant chimaera of the Pacific Coast of the United States and Canada. (*Bottom*) Head of a Spotted Ratfish. Chimaeras have distinctive lateral-line canals on their heads, highly reflective eyes, and mobile pectoral fins that they "flap" during swimming. Note also the beaklike teeth that give the group their alternative name, rabbitfish. Photos by Gene Helfman

The Giant Freshwater Whipray. This endangered species, which can reach a total length of 4.5 m (almost 15 ft) and a disc width of almost 2 m (6.5 ft), occurs in the rivers of Southeast Asia, New Guinea, and tropical Australia. Photo by Zeb Hogan; used with permission

An upper-jaw tooth from the extinct Megatooth Shark, with two teeth from a living White Shark shown for comparison. The Meg tooth measures 135 mm (5.5 in) on the slant. The largest Meg tooth found was 17.9 cm (7.25 in) long, suggesting an animal perhaps 17 m (55 ft) long. Wikimedia Commons, http://upload.wikimedia.org/wikipedia/commons/4/43/Megalodon_tooth_with_great_white_sharks_teeth-3-2.jpg

Owen Pyle inspects the jaws of a very large Tiger Shark captured south of the Hawaiian Islands. Tiger Sharks have an unmistakable tooth type and a highly varied diet and have been implicated in attacks on swimmers and surfers in the Hawaiian Islands and elsewhere. Photo courtesy of Richard Pyle and Lisa Privitera

This isn't an x-rayed ray but an embryonic Little Skate (*Leucoraja erinacea*) that has been "cleared and stained." The specimen is soaked in special dyes that stain different tissue types. Alcian blue binds to the cartilage in the skull, gill supports, backbone, and fin supports. Alizarin red binds to the enamel-hardened thorns concentrated along the tail and scattered around the body. The animal is then put in a series of chemicals that render muscle and skin clear. Clearing and staining is useful for revealing internal structures in studies of anatomy and taxonomy. Preparation and photo courtesy of Adam Summers

Sharks (0.2–53 km^2, or 0.08–20 square miles). Half of 36 tagged female Gray Reef Sharks in Palau remained within a few hundred meters of where they were tagged over a 2.5-year period, with occasional movements away from and back to their residence area. Such "site fidelity" is also considered a characteristic of Nurse, Blacktip Reef, and Caribbean sharks, especially among juveniles.

In Bimini, juvenile Lemon Sharks less than 1 m (3 ft) long initially patrol shallow beach areas of about 70 ha (0.7 km^2, or 0.3 square miles). This area increases with age: a 1.8-m-long shark may have a range of 18 km^2 (7 square miles), while a 2.3-m-long shark's home range encompasses as much as 93 km^2, or 36 square miles. Habitats also change with age, the shark moving from shallow nursery areas to open sand flats and finally to the reef and beyond. Whitetip Reef Sharks remain in small 1- or 2-km^2 areas (0.4–0.8 square mile) for several days up to six years, but they also periodically move between home range areas, sometimes traveling as far as 10 km between locales. Moving between different core areas appears to characterize several of the few sharks studied to date. An individual can therefore have several home ranges, staying for a while in one before moving to another a few kilometers apart, as is true of Blacktip Reef and Whitetip Reef sharks.

We know less about home ranges in other taxonomic groups. Cownose Rays are well known for their periodic long-distance migrations (see "Do sharks migrate?" in chapter 5). But some individual rays in Florida stay in fairly restricted areas for over a year. Average home range size for stay-at-homes was about 22 km^2 (8.5 square miles), but they spent 50% of their time in an area of only 3 km^2 (0.5 square mile).

Do sharks socialize with other kinds of animals?

Sharks don't enter into ecological relationships with many other animals, unless you count eating or being parasitized by them as a relationship. Some fishes follow sharks around, swimming just in front or under their host. Several members of the jack family Carangidae, such as juvenile Golden Trevally (*Gnathanodon speciosus*), do this. Pilot fish (*Naucrates ductor*) are seldom found away from a shark or manta ray host.

One family of bony fishes is highly adapted for associating very closely with sharks and rays. These are the remoras, or sharksuckers (family Echeneidae). Smaller sharks may have one or two small sharksuckers attached, but large species such as Whale Sharks and Manta Rays can have several large hitchhikers. Remoras attach not only to sharks and rays but also to marlins, whales, turtles, and an occasional diver. Remoras do this with an evolutionarily modified first dorsal fin that can create strong suction pressure. This allows the remora to remain attached to its host regardless of

how fast the host swims. A remora thus gets a free ride, covering distances that alone it would never be able to traverse, placing it in the right area when its host tears up prey. The remora then feeds on the scraps, as well as an occasional small nearby fish.

The costs and benefits of this relationship are a matter of debate. The shark may suffer a small energetic cost associated with the drag created by having a small hitchhiker. Some rays appear to find an attached remora annoying, which may explain why Mantas and Spotted Eagle rays jump out of the water. They could be trying to dislodge remoras when they land, but a definitive study hasn't been performed to test this idea. We do know that some sharks such as Caribbean Reef Sharks and Blacktip Sharks chafe (rub) their bodies against sand ripples to dislodge sharksuckers.

Recent findings also indicate that sharks and marlins may even benefit from an attending remora. Remoras apparently undertake cleaning behavior: their stomachs contain copepod parasites from the gills and fins of their hosts, making the relationship beneficial to both host and "parasite." Such a cleaning behavior relationship between an elasmobranch and another fish is not restricted to remoras and their hosts. Manta rays that have been bitten by sharks in Mozambique visit "cleaning stations" set up by various reef fishes such as wrasses. The mantas then hover in the water and let the cleaners pick over their wounds, removing dead tissue, attention that presumably prevents infection.

Another fish is often found in association with stingrays. This is the Cobia (*Rachycentron canadum*), a large bony fish (up to 68 kg, or 150 lb) that follows rays and some sharks around, presumably to feed on fishes and invertebrates flushed out of the sand or otherwise distracted by the foraging ray or shark. (Bottlenose dolphins off western Australia also follow large rays and steal prey such as octopus that the rays flush out of grass beds.) Fishers looking for Cobia home in on large rays and cast lures near them to catch the Cobia. Interestingly, Cobia are closely related to sharksuckers, suggesting that the relationship between sharksucker and shark may have first involved following behavior, which became more intimate as natural selection favored individuals that could attach themselves to a shark rather than having to expend energy staying close.

Sharks have benefited from associating with humans for probably as long as people have gone to sea. Large sharks such as Tiger and Blue sharks often followed sailing ships, feeding on scraps thrown overboard (not to mention an occasional luckless sailor). Sharks also learn that fishing activities create easy meals. This behavior goes beyond sharks hearing fish fighting on a fishing line, although anyone who has fished at sea, particularly in tropical waters, has experienced the disappointment of fighting a tuna for an extended period only to have the line go limp and find that only a head

is left to land. But some sharks have learned to follow wading anglers in anticipation of the activity. This behavior has been observed in several places in the Caribbean where anglers stalk Bonefish in shallow water, only to find they are in turn being followed by large Lemon Sharks. Bonefish are legendary for putting up a long and strenuous fight, after which the angler usually releases the Bonefish. But the long fight often exhausts the Bonefish, making it vulnerable to nearby predators. One study estimated that as many as 40% of released Bonefish are eaten by attendant Lemon Sharks.

Why do sharks swim at the surface with their dorsal fins sticking out?

Assuming we are talking about sharks in water deep enough to submerge their entire body, no one really knows the answer. This behavior certainly makes life easier for Hollywood filmmakers during dramatic, suspenseful moments, but that doesn't help the sharks much. The key probably lies in thinking more in terms of why sharks swim near the surface, which they certainly do more than many other large fishes. That their dorsal fins stick out is an artifact of having a stiff fin protruding from their back more than a result of the shark actually trying to stick it out of the water. One possibility is discussed in chapter 2 at the end of the section "How good is a shark's sense of smell?"

How smart are sharks?

In the several ways by which we measure intelligence—brain size, learning ability, behavioral complexity, creativity, or insightful behavior—we could be accused of stacking the deck in favor of humans, since we use measures on which humans tend to score high. These metrics may not necessarily be appropriate for assessing intelligence in other animals, especially sharks.

With those caveats in mind, scientists use two main types of information to assess intelligence in sharks: brain size and learning ability. Brain size has to be scaled relative to body size, because larger animals with bigger heads can obviously have larger brains, period. Two measures are therefore used, and argued about: brain weight relative to body weight, and the "encephalization quotient," a measure of brain size compared to an accepted standard (admittedly a human standard) (see "How large is a shark's brain?" in chapter 2). For present purposes, the first measure of brain weight relative to body weight will do.

It is difficult to determine evolutionary trends in brain size among elasmobranchs other than to say that groups that are more evolutionarily

advanced have larger brains. The relatively primitive squatiniform angel sharks and lamniform Basking Sharks have relatively small brains, as do squaliform dogfishes and rajid skates. White Sharks, also lamniforms, are intermediate. Carcharhinid requiem sharks, hammerheads, and dasyatid stingrays have relatively large brains, and odontaspid Sand Tigers have even larger brains. Mobuline manta and devil rays, which are generally considered the most advanced evolutionarily of all cartilaginous fishes, have the largest brains. Encephalization quotients move things around a bit, with freshwater stingrays moving to near the top of the list (see the next question, "Do sharks use tools?") and White Sharks slipping down. Regardless, by these measures, sharks have relatively large brains, at least when compared with bony fishes and even with higher vertebrates. The relative brain size of most sharks is larger than that of most bony fishes and even than that of about one-third of bird species and some mammals. Much of the increase is due to a relatively large and complex cerebellum.

What about learning ability? What we know is based on a limited sample size, involving lab tests on a few carcharhinids (particularly Lemon Sharks and Bull Sharks), orectolobids (Nurse Sharks), and small stingrays such as Round Stingrays and freshwater stingrays (Potamotrygonidae). We know of no such studies on chimaeras. In standardized learning studies, a shark receives a food reward if it learns a simple task, such as head-butting a button when hearing an underwater bell. Sharks do as well as bony fishes and mice in these tests and in tests of black versus white color discrimination. The sharks learn quickly and retain the knowledge, taking as much as six weeks to forget the task. Lemon Sharks learn quicker than Bull Sharks. No big deal.

How relevant are such tests to the field conditions and circumstances under which sharks have evolved? Put another way, are these valid tests of a shark's "intelligence" given the sorts of abilities that would be useful to a wild shark? The bell-head-butt-food-reward trials could be viewed as parallel to real world situations, such as a shark's learning that bony fishes make distress calls and erratic swimming noises when injured, a clue to a potential meal. In fact, later studies on newborn Lemon Sharks showed just this—that newborns don't react excitedly to prey distress sounds until they are "rewarded" with prey, after which they get excited when the sounds are played through underwater speakers. This study also showed that the sharks do better in a group, suggesting they may watch other individuals and learn from their behavior.

Other studies have shown that with experience, White-Spotted Bamboo Sharks become better foragers, in terms of being able to capture and consume novel prey more quickly. They also remember what they have learned, capturing prey efficiently even if they are allowed to forget for as

long as 18 days. Motoro Stingrays (*Potamotrygon motoro*) develop a mental map of their surroundings, learning the location of food items based on visible landmarks in their holding pens. This would certainly be a useful skill in the real world of a reef or a kelp bed, where various prey species tend to hide in certain places. It would also help them learn where divers will congregate with handouts during "interactive dives."

Do sharks use tools?

This is another one of those human-centered questions. We associate tool use with intelligence because we use tools and therefore wonder whether other animals are smart like us. Tool use implies taking an external object in the environment and using it to perform a useful task that otherwise would be difficult. There are a number of animal examples, such as a thrush (or a wrasse) banging a snail against a rock or a rock against a snail, or chimpanzees fishing termites out of a mound with a long blade of grass.

In this area, sharks pretty much fail, at least in terms of using external objects in food processing. By strict definitions, thresher sharks whacking prey with their tails, sawfish using their saws, stingrays displacing sand by flapping their wings or blowing water out their mouths so that they can excavate clams, or hammerhead sharks knocking stingrays to the bottom by delivering a downward blow with their "hammer" don't qualify as tool use. (See "How do sharks catch their prey?" in chapter 7.) One interesting variation involves Lesser Spotted Dogfish holding food items down with their tails. They apparently take advantage of the rough dermal denticles on the tail to hold the food in place while they tear pieces out of it with their teeth.

The example of rays excavating clams actually falls into a gray area because the ray isn't just using a body part to accomplish the task, but is instead pushing water with its pectoral fins or mouth. The water then becomes the tool, manipulated by the undulating fins or mouth muscles, to achieve a goal of obtaining food. In this context, researchers have studied Vermiculate River Stingrays (*Potamotrygon castexi*, also called the Otorongo Ray). Stingrays are potentially good subjects for investigation because batoids have the largest relative brains among elasmobranchs.

The study looked not only at tool use but also at learning ability in a problem-solving situation. The researchers placed food rewards inside an opaque plastic tube that had a mesh screen divider, which meant the food could be obtained only if the ray tried to extract the food from the "correct" end. (Such "tube-traps" are commonly used in studies of animal learning.) All five test animals improved their performance by figuring out progressively more quickly with each trial which end of the tube to work from, correcting themselves if they first tried to extract the food from the

end that was blocked by the mesh. Each ray "rapidly learned to use water as a tool to extract food from the testing apparatus." Interestingly, all five started out using only one method, either creating flow by undulating their pectoral fins or creating suction by "capping the pipe" and flexing their body. By the end of the trials, they were all using both tactics. So they corrected for their mistakes and adjusted their methods, all of which led to obtaining the food faster.

In defense of selachian sharks, we would offer that evolution has provided them, unlike us, with all the "tools" they need for normal activities.

Do sharks play?

Play must have evolved after sharks came on the scene. With a few possible exceptions, we are unaware of convincing examples of playlike behavior in sharks. But it's a big ocean.

Some behavior by White Sharks has been interpreted as object play. Boat crews in South Africa have suspended cloth bags with bait. Known individual White Sharks grabbed hold of the bag while the crew pulled on the other end in a sort of tug-of-war, like a dog pulling on the other end of a sock. The sharks showed no indication that they were trying to eat the bag or its contents. These sharks also pushed floats around at the surface, with or without bait present. Is this play or something else? You be the judge.

Another, more intriguing possibility comes from the Porbeagle, a relative of the White Shark. Several websites describe apparent play behavior observed in groups of Porbeagles off the Cornish coast of England. They are said to roll over and wrap themselves in kelp, to chase other Porbeagles that have kelp draped over them and bite at the kelp, and to "play" with objects at the surface such as pieces of driftwood or fishing floats, including "passing objects from one individual to another and repeatedly tossing the objects clear of the water." We have been unable to find the original reports of this behavior, and the web reports, all of which are secondhand, appear to repeat one another.

Do sharks talk?

Sharks, unlike bony fishes, are not known to make sounds. Bony fishes rub bones together, grind throat teeth, or vibrate their gas-filled swim bladders. Sharks have neither bones nor gas-filled bladders; grinding teeth would be possible only in rays or maybe horn sharks. Other than incidental sounds associated with crushing prey, sharks maintain silence (despite the roaring sharks that seem to populate many of the grade B shark attack

movies). However, sharks do locate bony fishes by eavesdropping on their prey's sounds.

What is "tonic immobility"?

If you read popular articles about sharks or watch Discovery Channel specials or YouTube videos, you're likely to encounter instances of tonic immobility (TI). For reasons that remain unknown (as with so many other aspects of shark behavior), when some sharks are turned over and held upside down, or when their snout is pushed up and backward, or just squeezed gently, the shark goes into something of a trance and stops moving. One untested possibility is that a shark's electroreceptors are concentrated in the snout area, and perhaps overstimulating this area "shocks" them into immobility (granted, this does not explain the reaction to being turned upside down). Whatever the explanation, researchers have used this reaction to calm sharks that they want to measure or tag because it makes things easier on both shark and researcher. Sharks in TI are less likely to injure themselves, equipment, or people, and they recover from TI faster than if an anesthetic is sprayed across their gills. See www.youtube.com/watch _popup?v=WK2LpUoqX6A&vq=medium for a nice example set to music, or watch a couple of scenes in the otherwise forgettable Hollywood blockbuster *Dark Tide*.

As a side note, many of the photos showing White Sharks lunging out of the water with their mouths open are not instances of foiled attacks. These are instead sharks that have been lured to the surface with bait and had someone grab and push hard against the underside of the snout, inducing a short-lived TI state, long enough to get the photo. We do not recommend that you try this at home.

Chapter 5

Shark Ecology

Do sharks migrate?

Although some sharks, such as Bull and Bonnetheads, are comparative stay-at-homes (see "Are any sharks territorial?" in chapter 4), others travel long distances on a regular basis. It isn't surprising that many long-distance movers (Whale Sharks, Basking Sharks) are large, but even some relatively small sharks such as Spiny Dogfish will cross entire ocean basins. Migratory behavior is in fact a prime reason that many sharks are overfished.

Migrations can be both horizontal and vertical. Horizontal movements are generally much greater and can be along coastlines or across oceans. Coastal species stay near continental shelves, over depths of generally less than 200 m (650 ft), but can still travel great distances. Coastal sharks include threshers, Sandbar, Blacktip, and Dusky sharks. Bigeye Threshers (*Alopias superciliosus*) have been tracked from waters off New York to the eastern Gulf of Mexico, a distance of 2,767 km (1,719 miles). Sandbar Sharks move between the northeastern United States and the Yucatan Peninsula of Mexico, a distance of about 5,600 km (3,480 miles). Great Hammerheads are found along coastlines and also far at sea over seamounts. One "coastal" individual was followed for 62 days. It left South Florida and traveled north to New Jersey, sometimes drifting with the Gulf Stream. It traveled a minimum distance of 1,200 km (745 miles). Scalloped Hammerheads tagged at Malpelo Island, Colombia, have been followed to the Cocos Islands off Costa Rica (627 km, or 389 miles); and one shark then moved to the Galapagos Islands, another 710 km (440 miles), for a minimum distance of 1,340 km. Because Malpelo, Cocos Island, and the Galapagos are among the few places where Scalloped Hammerheads are known

A Scalloped Hammerhead at Cocos Island, off the coast of Costa Rica. Scalloped Hammerheads migrate between Cocos Island, the Galapagos Islands, and Malpelo Island off Colombia, forming large aggregations. To protect this globally endangered species, fishing in all three areas needs to be stopped. Photo by Barry Peters, Wikimedia Commons, http://en.wikipedia.org/wiki/File:Scalloped_hammerhead_cocos.jpg

to congregate (see "How can I see sharks in the wild?" in chapter 8), it is likely that some of these are the same sharks, meaning hammerhead populations are smaller than if we were looking at entirely separate groups.

Even comparatively small coastal sharks can cover large distances during their lifetimes, and the distinction between "coastal" and "oceanic" is far from hard and fast. Some coastal species will cross deep ocean areas to get to another coast. A School Shark less than 1.5 m (5 ft) long that was tagged off New Zealand's South Island was caught 4,940 km (3,069 miles) away off South Australia, 3.5 years later. To get to Australia, it had to cross the stormy Tasman Sea, which is more than 5,180 m (17,000 feet) deep. Spiny Dogfish are most common along coastlines and generally move up and down a coast, but some fish don't know this. One tagged off the coast of Newfoundland crossed the Atlantic Ocean and was captured near the Shetland Islands 11 years later, a minimum distance of more than 3,200 km (2,000 miles). Spiny Dogfish tagged off the west coast of Canada moved long distances along the coast, traveling north in spring and summer and south in fall and winter. Some were recaptured years later up in Alaska and south in Washington State, Oregon, California, and Mexico. But several traveled to Japan, across the North Pacific Ocean.

The champion horizontal migrators are species that cross ocean basins or move north and south between hemispheres. Blue Sharks make round trips across the Atlantic between North America and Europe, a straight-line journey that exceeds 16,000 km (9,900 miles), although the sharks undoubtedly cover more area because it's unlikely they swim in a straight line. Some Blue Sharks move north and south; a Blue Shark that was tagged off New York was later recaptured off Brazil, a one-way straight-line journey of 5,980 km (3,715 miles). Some other long distances traveled by oceanic species in the North Atlantic include 6,700 km (Tiger Shark), 4,500 km (mako), and 2,800 km (Oceanic Whitetip)—but again, these distances

are underestimates because they record only starting and ending points for sharks tagged at one place and recaptured at another.

White Sharks are now known to migrate between central California and Hawaii, a minimum one-way distance of 3,800 kilometers (2,360 miles); the return trip makes the journey almost 8,000 km (4,970 miles). Some "Californian" White Sharks move back and forth along a fairly narrow corridor to a place in the Pacific about 2,500 km (1,550 miles) west-southwest, taking only three to four weeks to make the journey. They spend fall and part of the winter in California and most of spring and summer in the offshore area, now called the White Shark Cafe. White Sharks tagged off South Africa go to western Australia and back. One 4-m (13-ft) female nicknamed Nicole (after Australian actress Nicole Kidman, an advocate of shark conservation) was followed by satellites across the Indian Ocean and then back, traveling between South Africa and Australia. The round trip took nine months, covering a minimum distance of 22,000 km (13,670 miles).

Whale Sharks move around and across oceans, arriving at particular places to feed on eggs spawned by reef fishes and invertebrates. A Whale Shark tagged off Mexico traveled 13,000 km (8,000 miles) across the North Pacific Ocean. Another tagged in the Philippines traveled to Vietnam, a distance of 4,567 km (2,830 miles), in only 2.5 months, while another tagged off Malaysia moved 8,025 km (4,987 miles) over a 4-month period. Many of these movements involve migrations to areas of briefly abundant food; after the food is gone, the animals may return to their starting point or go to places that we don't know about yet. Despite all this moving around, genetic analysis of sharks sampled from the eastern Pacific (Mexico, Costa Rica, Galapagos Islands), the Caribbean and the western Atlantic (Honduras, Florida), and South Africa, western Australia, and the Indian Ocean (Seychelles, Maldives, Djibouti, India) indicates that sharks intermingle often when breeding. Basically, there is one world population of Whale Sharks totaling between only 250,000 and 500,000 individuals, a finding with strong conservation implications. (For comparison, the Spiny Dogfish population off Canada's east coast may number over 500 million sharks; see "Which sharks are most or least abundant?" below.)

Basking Sharks, second in size only to Whale Sharks, also migrate across oceans. Baskers tagged off the United Kingdom have been found three months later off Newfoundland, Canada, on the other side of the North Atlantic. Sharks tagged off the Cape Cod area during summer headed south as water temperatures fell in autumn. Some went to the Bahamas, others to the Caribbean, and a few traveled as far south as Brazil, a straight-line distance of 6,500 km (4,030 miles); the actual track of one shark indicated it had moved nearly 9,000 km (5,580 miles). While in productive northern areas, Basking Sharks feed at the surface, but when in the

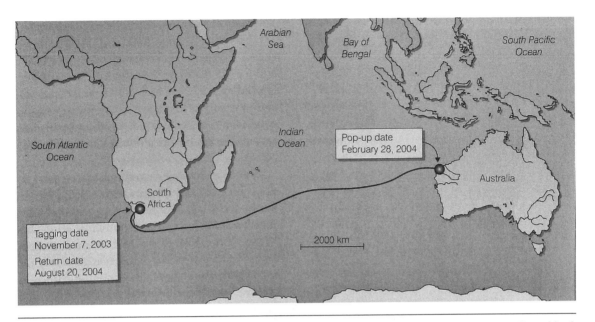

The migration track of Nicole, a 4-m female White Shark who swam from South Africa to western Australia and back over a nine-month period. Redrawn from R. Bonfil et al., "Transoceanic migration, spatial dynamics, and population linkages of white sharks," *Science* 310 (2005): 100–103; used with permission of Wiley-Blackwell

tropics, they descend to open ocean depths as great as 1,000 m (3,300 ft), where they remain active for months, probably feeding.

These findings solve one of shark biology's great mysteries, namely where Basking Sharks go in the winter. Because they "disappeared," it was guessed that they hibernated on the bottom somewhere. The new information is based on "pop-up" tags that periodically record depth, temperature, and location: these tags break free from the shark, float to the surface, and transmit their stored data via satellite to a laboratory (see "How do scientists study the movements of large sharks?" in chapter 12). We don't know what the sharks do after their tags are shed. It is assumed that they return to northern areas the following spring because that is when they start to reappear in surface waters off New England. Learning that Basking Sharks probably feed in deep, cool waters for months at a time also calls into question their name. They are called "Basking" Sharks because it was assumed that their slow movement at the surface was for sunbathing (Irish fishermen used to also call them "sunfish"). It now seems more likely that they are at the surface because that's where their food is located at high latitudes. We are open to suggestions for a new name for this plankton-eating, migrating giant.

We know less about ray migrations, especially among smaller species. Atlantic Stingrays migrate seasonally, apparently avoiding cold water. During summer and fall they occur in areas such as the Chesapeake Bay, but when surface waters cool, they move south or offshore to warmer water.

Shark Ecology

Spotted Eagle Rays in the Bahamas move offshore in early summer and return to shallow reef areas by late summer. Manta ray behavior is slowly being revealed; they too migrate. For example, Giant Mantas migrate across the Indian Ocean from Mozambique to the Maldives, a distance of 1,120 km (700 miles) in two months. Cownose Rays are among the better-known species because they migrate in large numbers, often near the surface, and are therefore visible from boats and planes. Cownose Rays move seasonally along the U.S. Atlantic coast, going north in the late spring and south in late fall between Virginia and Florida and even as far as between the Chesapeake Bay and Brazil.

A few studies have looked at whether skates migrate. Because they are small animals that lie on the bottom in deep, cool water, we wouldn't expect them to be highly mobile, and many aren't. A 20-year study of Thorny Skates off Newfoundland found that most skates didn't undergo obvious seasonal migrations in any particular direction but instead moved slowly from their tagging location. Many skates moved less than 30 miles (48 km), and most less than 60 miles (96 km), even after many years. One moved less than 5 miles (8 km) in 14 years. Among five skate species that occur along the Atlantic coast of the United States (Thorny, Smooth, Clearnose, Little, and Winter), Thorny and Clearnose skates did not migrate, whereas Clearnose, Little, and Winter skates undertook apparent seasonal migrations. Clearnose Skates move both north and south as much as 580 km (360 miles), and inshore and offshore as much as 120 km (74 miles). The inshore movement in early spring is probably a spawning migration. Some Little Skates move inshore in the spring and offshore in autumn and winter. Winter Skates are the strongest migrators, although again not all individuals move. In winter and spring, Winter Skates move south or offshore from an area in the Gulf of Maine, some animals traveling as much as 940 km (584 miles). But compared with their shark and ray relatives, these skates are weak migrators.

The movements of chimaeras are poorly understood. The Australian Ghostfish of Australia and New Zealand normally occurs in fairly deep water, down to 200 m (656 ft). Reproductively active animals move into shallow bays to lay eggs during their summer and autumn spawning season. They move back offshore to deep water in winter and spring. Egg cases of the Elephantfish, *Callorhinchus callorynchus*, have also been found in shallow water, suggesting that this chimaera also moves inshore to spawn. The Spotted Ratfish is a vertical migrator, moving up from deeper water in the day to shallower water at night. Aside from these three, we can't say much about chimaera movements.

Many sharks move up and down periodically or irregularly. Although the distances traveled are less spectacular than transoceanic movements,

the vertical migrations of some sharks are impressive. Vertical migrators experience greater changes in temperature and pressure over a few hours than do sharks moving the length of a coastline or across an ocean over many days. The diving profiles of these sharks, as revealed by ultrasonic telemetry and archival tags, show a yo-yo pattern: animals move up and down several hundred meters repeatedly during the day and night or over the course of several days. Tiger Sharks have been described as "bounce divers," moving almost continuously between the surface and 90 m (300 ft) down. White Sharks may move up and down between the surface and 200-m (660-ft) depths as often as 96 times in 24 hours; others may spend days at the surface and then more days at depth. Female Gray Reef Sharks observed in Palau exhibited a distinctive, complex up-and-down move-ment pattern that extended over only 15 m (50 ft) each day. They tended to move to shallower water (30 m, or 100 ft, deep) at dawn and dusk but spent daylight hours at 45-m (150-ft) depths. The depths at which they stayed at night were usually shallower than daytime depths, except on full moon nights, when they descended to 60 m (200 ft).

Bigeye Threshers, Megamouths, Shortfin Makos, and School Sharks spend daytime at depths of several hundred meters and then move to within 10 to 100 m (33–330 ft) of the surface for the night, probably tracking food. Whale Sharks engage in both irregular and day versus night vertical move-ments, moving between the surface and 1,000 m (3,300 ft). They spend most of their time near the surface regardless of light, especially when feeding. Their attraction to the surface is probably because their food—recently spawned fish and invertebrate eggs—is concentrated in shallow water. Basking Sharks around the British Isles move up and down in the classic pattern of going deeper (to 70–80 m, 225–265 ft) by day and staying shallower (10–30 m, 33–100 ft) at night when they are in deep water that is colder at depth. But in shallower areas where there is no temperature dif-ference with depth, these sharks stay in shallower water by day and deeper water at night. In both the classical daytime-deep/nighttime-shallow and reversed patterns, it appears the Basking Sharks are following the move-ment of their zooplankton prey.

A few exceptional sharks undertake a daily vertical migration clearly linked to food availability. Many bony fishes (lanternfishes, lampfishes, bristlemouths, lightfishes) live at mesopelagic (middle-ocean) depths be-tween 200 and 1,000 m (660 and 3,300 ft) during daylight hours. They move up near the surface in the evening to feed under the cover of dark-ness on the zooplankton that is more abundant in the shallows. They are met there by a number of predators, such as dolphins, billfishes, tunas, and Megamouth Sharks. These predators find their prey in the dark because mesopelagic fishes are often bioluminescent, giving off a green glow from

light organs on their sides and bellies. Capitalizing on this predator-prey interaction are Cookiecutter Sharks. Cookiecutters are also bioluminescent, but for devious reasons. The Cookiecutters hide among the mesopelagic fishes, apparently blending into the bioluminescent background, and ambush their prey. Cookiecutters aren't feeding on zooplankton or mesopelagic fishes but on the predators of the vertical migrators. Unlike most sharks that chop up their prey and swallow large mouthfuls, Cookiecutters feed one small bite at a time, taking a single, circular, half-dollar-size plug out of their victims' sides. (See "What kind of teeth do sharks have?" in chapter 2 and "Do sharks chew their food?" in chapter 7.)

Because many sharks are highly migratory species, moving around and through entire ocean basins, they encounter fishing boats from many countries. This exposes them to capture, repeatedly. A shark may be protected by local or national fishing regulations when near the shoreline, but the high seas are largely unregulated, and few police exist to enforce any laws on the books. As a result, pelagic (open-ocean) sharks are heavily overfished (see chapter 10).

How do sharks navigate?

Moving across an ocean and back again, returning to the exact same locale after a nine-month absence, as was the case with Nicole the White Shark, requires a highly sophisticated navigational system. So does arriving at the right place and time to feed on prey that is available for only a short period. Tiger Sharks show up in the northwestern Hawaiian islands just as young Blackfooted and Laysan albatross are taking their first, sometimes unsuccessful and therefore fatal, flights. Tiger Sharks do the same thing at Raine Island, Great Barrier Reef, Australia, just as green sea turtles appear by the thousands to lay eggs. At Raine Island, the Tigers scavenge heavily on dead turtles that had been stranded at low tide on exposed reefs. Some Tiger Sharks occur year-round at both locations, suggesting that Tiger Shark populations consist of resident and transient or migratory individuals. Whale Sharks are also famous for showing up at different locales around the tropics just when reef fishes or invertebrates are about to spawn. Manta rays do the same. White Sharks arrive at the Farallon Islands off San Francisco in late autumn to feed on young elephant seals that arrive there at about that time.

How sharks accomplish these and other precise long-distance navigational feats remains a mystery. Our best guess is that they rely on their acute electric sense, navigating via geomagnetic cues related to the earth's magnetic field and its anomalies and gradients (see "How do sharks detect electric fields?" in chapter 2). This kind of navigation means that sharks

have an internal (genetic or learned) map, compass, and clock. They compare where they are and in what direction they are moving with that mental map. The details of this ability remain one of the great mysteries of shark biology. Someday, a shark biologist, perhaps a reader of this book, will discover the answer.

How many shark species live in rivers and lakes?

We think of sharks as marine animals, and this is mostly true. Of the 1,250 or so known species of sharks, skates, and rays, fewer than 50 live all or most of their lives in fresh water. (Chimaeras live only in the ocean, usually far from coastlines.) Among the "freshwater" elasmobranchs are one guitarfish, 6 sawfishes, 8 sharks, and 31 stingrays.

Stingrays make up the largest group of freshwater elasmobranchs. Two stingray families enter or live full-time in rivers and lakes. The largest family, members of which are entirely freshwater inhabitants, consists of the potamotrygonid river stingrays, with 20 species, 18 of which are in the genus *Potamotrygon*. All live in South America, particularly in the Amazon River. Many are popular aquarium pets, despite the strength of the poisons in their barbs. The other group consists of 11 species in the large stingray family Dasyatidae. Most freshwater dasyatids occur in Africa, Southeast Asia, India, Myanmar, northern Australia, and New Guinea. This family includes what may be the world's largest stingray, the Giant Freshwater Stingray of Southeast Asia. These rays can be 4.6 m (15 ft) long and 1.9 m (6.2 ft) across the back and can weigh more than 500 kg (1,100 lb). One population of Atlantic Stingrays lives in the St. Johns River of Florida, more than 350 km (217 miles) from the coast. This makes it the only elasmobranch in North America that spends its entire life cycle in fresh water. These rays are frequently observed by snorkelers in the crystal clear springs that flow into the St. Johns River.

Lemon Sharks occasionally enter brackish river mouths, but we don't normally think of them as "freshwater" sharks. Five other species of carcharhinids, all in the genus *Glyphis* and all residents of Asian or Australian rivers, are permanent freshwater or estuary dwellers. Their names suggest their habitats: the Ganges Shark of India (*Glyphis gangeticus*), the Northern River Shark of New Guinea (*G. garricki*), and the Irawaddy River Shark of Myanmar (*G. siamensis*). Undoubtedly the most infamous freshwater shark is the Bull Shark. Most Bull Sharks live along coastlines and are perfectly good marine inhabitants, aside from their well-deserved reputation for attacking humans (see "Are sharks dangerous to people or pets?" in chapter 9).

Bull Sharks, however, commonly swim up rivers, often quite far, and are known to have attacked people in rivers in Iran, India, and South Africa. In

The Alton, Illinois, Bull Shark. In 1937, something was tearing up the wooden fish traps set by these two fishermen, Herbert Cope and Dudge Collins, so they built a large trap out of wire mesh and baited it with chicken guts. To their surprise, the culprit was a 1.5-m (5-ft), 38-kg (84-lb) Bull Shark. The shark would have traveled a minimum of 1,160 river miles from the mouth of the Mississippi to get to Alton. Photo by Ralph Manns

1937, two commercial fishermen in Alton, Illinois, were surprised to find a 1.5-m-long (5-ft) Bull Shark in a trap baited with chicken guts. Alton lies about 1,160 river miles (1,850 km) from the mouth of the Mississippi and 15 miles (24 km) north of St. Louis, near where the Mississippi, Illinois, and Missouri rivers come together. Alton declares itself "the most haunted city in America" because of the many ghosts that are claimed to live there (including one Bull Shark?). Today, a shark trying to get to Alton would have to traverse a lock and dam built in 1953, so it's unlikely any have made the trip since then (three juvenile Bull Sharks reportedly caught in Wisconsin and Minnesota in 2006 were part of an April Fools hoax).

Bull Sharks have also been found as far as 4,200 km (2,610 miles) up the Amazon, and they regularly traverse the 175-km-long (105-mile) Rio San Juan of Nicaragua, moving between the Caribbean Sea and Lake Nicaragua, where they are called Lake Nicaragua sharks. They also move up other rivers in Mexico, and Central and South America. Bull Sharks, along

Sharks: The Animal Answer Guide

Sawfishes were once common in tropical nearshore areas around the world. They are now among the world's most imperiled marine and estuarine fishes. Shown here are saws taken from sawfish during a 1920 fishing tournament in Key West, Florida. Photo property of Matthew McDavitt; photographer unknown

with Largetooth Sawfish (*Pristis microdon*), have been heavily overfished in Lake Nicaragua and are now protected by Nicaraguan law. Bull Sharks occasionally swim up a river and then get stranded in water bodies after a flood. This is the probable explanation for how several Bull Sharks got into a water hazard at the Carbrook Golf Club in Brisbane, Australia. It's prudent to just take a penalty stroke if you find yourself in the lake on the 15th fairway at Carbrook.

All six (or maybe seven) species of sawfishes inhabit estuaries and rivers. The Largetooth Sawfish has been found as far up the Amazon as Santarem, Brazil, more than 800 km (500 miles) from the sea. Largetooth Sawfish established a genetically distinct reproducing population in Lake Nicaragua but have become increasingly rare as a result of overfishing and habitat destruction. Because of their large size, dangerous tooth-studded snouts, and tremendous strength, sawfishes are the stuff of legend and respect in many countries (see "What roles do sharks play in religion, mythology, and legends?" in chapter 11). Unfortunately, they are also among the most endangered elasmobranch species in the world.

Several sharks and sawfishes move freely between marine and fresh water, prompting the question of how they deal with such drastically changing salinities. Unlike the full-time freshwater species, these estuarine migrants can adjust the molecular concentrations of their blood. A Bull Shark in fresh water reduces the salt concentration of its blood by about 20% and the urea concentration by about 50%. It also compensates for the extra

water that comes into its body by increasing its urine production 20-fold. These and other physiological adjustments are quickly reversed when the shark returns to the ocean.

People live, work, fish, play, and dump their trash in rivers. As a result, freshwater elasmobranchs are more threatened than are chondrichthyans in general (see "Are any sharks endangered?" in chapter 10).

How far down in the ocean do sharks live?

Sharks do not live as deep in the ocean as bony fishes. The deepest living bony fish is a cusk eel, *Abyssobrotula galatheae*, found at 8,370 m (27,460 ft). The depth records for sharks are held by a Great Lanternshark (*Etmopterus princeps*) at 4,500 m (14,763 ft), a Portuguese Dogfish at 3,675 m (12,057 ft), and a Leafscale Gulper Shark (*Centrophorus squamosus*) at 3,280 m (10,761 ft), suggesting that sharks go only half as deep as bony fishes. The record for skates is held by Bigelow's Skate (*Rajella bigelowi*), at 4,156 m (13,635 ft), and the Pallid Skate (*Bathyraja pallida*), at 3,280 m (10,761 ft). Holocephalans appear to bottom out around 3,000 m (9,900 ft), represented by the Atlantic Chimaera (*Hydrolagus affinis*) at 2,909 m (9,544 ft) and an unidentified member of the genus *Harriota* at 3,010 m (9,875 ft).

Chondrichthyan fishes as a group may avoid greater depths because of their comparatively large size, which would be hard to maintain given the general lack of food in the huge expanse of the ocean's deepest waters.

This is not to say that sharks don't live in fairly deep water. In fact, about half of all shark, ray, and chimaera species occur predominantly in water deeper than 200 m (660 ft), which is near the cutoff depth that marine biologists use to define the deep sea. Sharks occur on and just above the bottom as well as in open water. Breaking deepwater sharks down by groups, 254 sharks (53% of all sharks), 236 batoids (38%), and 40 holocephalans (89%) occur below 200 m. Twenty-three families of sharks appear in deep water, but of these, only five families contain more than 10 species: scyliorhinid catsharks (102 species), etmopterid lanternsharks (42 species), squalid dogfish sharks (25 species), centrophorid gulper sharks (18 species), and somniosid sleeper sharks (16 species). The remaining families contain only one or a few deep-sea members. Some of the world's strangest sharks are deep-sea denizens, with several monotypic (only one species in the family) groups represented. These include the Frilled Shark (*Chlamydoselachus anguineus*), the Crocodile Shark (*Pseudocarcharias kamoharai*), the Goblin Shark (*Mitsukurina owstoni*), and Bramble and Prickly sharks (*Echinorhinus brucus* and *E. cookei*). Carcharhinid sharks dominate shallow seas, but only two species in this family, the Bignose Shark (*Carcharhinus altimus*) and the Night Shark (*C. signatus*), spend much time in the deep.

Sharks: The Animal Answer Guide

Skates live on the sea floor, with three families—rajid hardnose skates, arhynchobatid softnose skates, and anacanthobatid legskates—making up 90% of the 236 deep-sea skate and ray species. Only five species in four families of stingrays occur in the deep sea, of which two are monotypic: the Sixgill Stingray (*Hexatrygon bickelli*) and the Deepwater Stingray (*Plesiobatis daviesi*). As with the carcharhinid sharks, the whiptail stingray family Dasyatidae, so successful in warm, shallow waters, has only one deep-sea member, the Shorttail Stingray, which also occurs in shallow water. Holocephalans are well represented in the deep sea; 40 of the world's 46 chimaera species are deep-sea animals. As a group, chimaeras generally swim just above the bottom.

A few of the deep-sea sharks are large: somniosid sleeper sharks can be 6 m (20 ft) long, and hexanchid sevengill and sixgill sharks grow to 3 m and 5 m (10 ft and 16.5 ft), respectively. Otherwise, deep-sea sharks are generally small and include the world's smallest sharks, maturing at only 17 to 25 cm (7–11 in) in length, much smaller than the vast majority of shallow-water sharks. These microsharks include the etmopterid lanternsharks, dalatiid pygmy sharks, and proscylliid finback catsharks. (See "What are the largest and smallest sharks alive today?" in chapter 1.)

Which geographic regions have the most species of sharks?

Zoologists divide the world's oceans into four major zones, depending mostly on temperature: tropical, warm temperate, cold temperate, and polar. Each zone is further divided by oceans and continents, and oceans are again divided according to depth.

Overall, the warm temperate regions house the greatest combined diversity of chondrichthyans (720 species). The tropics come next, with 654 combined species. Then come cold temperate regions (296 species); and last are polar seas, with only six species, all in the north polar (Arctic) zone. Many species, such as Tiger Sharks and Spiny Dogfish, occur in several areas and are thus counted more than once in this distribution, so the numbers far exceed the world's overall shark diversity of 520 or so species.

Exploring further, the North and South Pacific warm temperate regions combined have more species than the North and South Atlantic regions (340 versus 292 species), with a small region of high diversity around South Africa (88 species). The Indo-West Pacific tropical area (which encompasses the western tropical Pacific plus Indian oceans) is home to about 1.5 times more species than the tropical eastern Pacific and tropical Atlantic areas combined (423 versus 231 species). In the cold temperate seas, the highest diversity is around Australia and New Zealand (128 species),

followed by the North and South Atlantic (92) and the North and South Pacific (26). Another hotspot of 40 cold temperate sharks is found around the southern tip of South America, but true sharks are absent farther south around Antarctica.

The diversity of the major taxonomic groups varies by geographic region. The greater overall shark diversity in warm temperate regions is in part due to the large number of squaliform dogfish shark species there (200 species); squaliform sharks are also abundant in the tropics (143 species) and in cold temperate waters (100 species) but not in the Arctic (5 species). Scyliorhinid catsharks are likewise (slightly) more diverse in warm temperate than tropical oceans (97 versus 83 species), falling off in cold temperate waters with only 33 species, and no polar species. Carcharhinid requiem sharks, in contrast, are most diverse in the tropics (184 species) and the warm temperate regions (169 species), with a remaining 48 cold temperate carcharinids. Orectolobiform carpet, bamboo, and Nurse Sharks are another "majority tropical" group, with 44 tropical species, 33 warm temperate species, and 10 cold temperate representatives.

Comparing distribution by depths, squaliforms are most diverse in the deep sea, carcharhinids are coastal and oceanic, scyliorhinids are coastal and deep sea, and orectolobiforms are sharks of coastal shallows.

The batoids (electric rays, sawfishes, guitarfishes, skates, and stingrays) are primarily tropical and warm temperate elasmobranchs that live on the bottom in relatively shallow water—though with many exceptions. Torpediniform electric rays occur worldwide in shallow warm temperate and tropical oceans, with a few found as deep as 1,100 m (3,630 ft). Pristid sawfishes are nearshore, shallow warm temperate and tropical fishes, some of which enter fresh water. Rhinobatid guitarfishes are also worldwide in primarily shallow warm temperate and tropical bottom regions. The rajid skates have a much wider geographic and depth distribution, ranging from the poles to the equator and as deep as 3,000 m (9,900 ft). As a group, they tend to live in deeper, cooler water and at higher latitudes than the other batoids, including a species that occurs in Antarctica and nowhere else. The myliobatiform rays are again found worldwide in tropical to warm temperate seas, primarily in shallow water, with a number of tropical freshwater dasyatoid stingrays such as the potamotrygonid freshwater rays of South America. Overall, the batoids are most diverse in the Indo-West Pacific region. North and South America are second in batoid diversity.

The holocephalan chimaeras inhabit all but polar seas. Most are deepwater dwellers at 500 m (1,640 ft) and deeper. One family, the Callorhynchidae, occurs only in the Southern Hemisphere (South America, South Africa, and Australia/New Zealand). Members of the other two families (Rhinochimaeridae and Chimaeridae) are found in all the world's oceans.

Some species occur throughout an ocean basin; this is the case with the Pacific Spookfish (*Rhinochimaera pacifica*) in the Pacific Ocean and the Broadnose Chimaera (*R. atlantica*) in the Atlantic. The Narrownose Chimaera (*Harriotta raleighana*) occurs in both the Pacific and the Atlantic. Australia and New Zealand may have the highest diversity of chimaeras, with at least nine species recorded from that region.

Which sharks are most or least abundant?

Probably the most abundant shark in the world is the Spiny Dogfish (also known as the Spur Dog or Piked Dogfish), *Squalus acanthias*, and its close relative, *Squalus suckleyi*. *S. acanthias* occurs off the Atlantic coasts of the United States, Canada, and Europe, and also off southern South America, Africa, Australia, and New Zealand. The closely related *S. suckleyi*, until recently considered the same species, inhabits the North Pacific. Commercial landings of the latter off the west coasts of the United States and Canada exceeded 50,000 metric tons (100 million pounds) per year in the mid-1940s. At about 3 pounds average per fish, that is over 33 million sharks! Some recent estimates of the abundance of the Atlantic species off the coast of Canada run as high as 535 million sharks, and the number off the combined Pacific coasts of Canada and the United States is put at almost 2 billion individuals. As abundant as those numbers appear, these

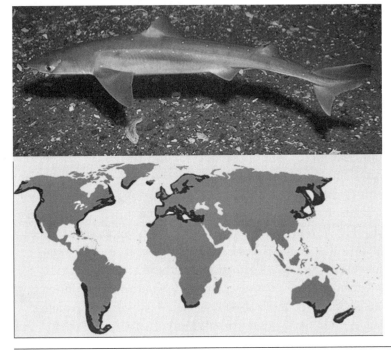

(*Top*) A Spiny Dogfish, probably the world's most abundant shark, off the coast of New England. The typical color is bronze-gray with white spots. (*Bottom*) Geographic ranges, shown in black, of the Spiny Dogfishes *Squalus acanthias* and *S. suckleyi*. The latter occupies the North Pacific. *Top*, photo by Doug Costa, National Oceanic and Atmospheric Agency, https://marinelife.noaa.gov/media_lib/preview.aspx?ID=7299&p=img; *bottom*, map from http://en.wikipedia.org/wiki/File:Squalus_acanthias_distmap.png

are still sharks with low reproduction rates (see chapter 6). Some dogfish stocks have been fished down as much as 95%, leading to restricted catches and a "Do Right by Dogfish" campaign to reduce overexploitation.

Some other abundant sharks include Blue, Silky (*Carcharhinus falciformis*), School or Tope, and Gummy (*Mustelus antarcticus*) sharks—all of which are targets of major fisheries. Among batoids, Bat and Cownose rays occur in schools of thousands and even millions of individuals (see "Do sharks form schools?" in chapter 4). Spotted Ratfish may be the most abundant chimaera, with an estimated 200 million in Washington State's Puget Sound, an area of about 2,600 km^2 (1,000 square miles). That's roughly 77,000 ratfish per square kilometer, or about one ratfish every 10 m^2 (equivalent to an area of 30 by 30 ft).

It's hard to say what the rarest shark, skate, ray, or chimaera might be. The rarer species are certain small sharks and rays that are endemics (confined to a very small geographic region). Many elasmobranchs are threatened from overfishing and have declined greatly in numbers (see chapter 10). Sawfishes as a group are among the most depleted because they live near shores and in estuaries, habitats that put them in contact with people. Even when they are not targeted by fisheries, their saw can become entangled in any kind of net, and they drown or are killed so that someone can claim the saw as a trophy (even though possession of one is illegal).

Some of the least known sharks, such as small deepwater species, may or may not be rare; we don't know for sure because they are difficult to capture or live in places that are poorly explored or hard to get to. These include some surprisingly large species, such as the Goblin Shark, which can grow to 3.3 m (11 ft) long and weigh 159 kg (350 lb); fewer than 50 specimens are housed in museum collections. Among rare large sharks, the 5.5-m-long (18-ft) Megamouth Shark gets the most attention because it wasn't discovered until 1976 and, as of 2012, only 54 individuals have been seen.

Several sharks are known from only one or a few specimens that are housed in museum collections. The Irrawaddy River Shark, a carcharhinid, is known from a single museum specimen that was caught in the late 1890s at the mouth of the Irrawaddy River in Myanmar. A close relative, the Northern River Shark, may have a total global population of only 250 mature individuals. The Pondicherry Shark (*Carcharhinus hemiodon*) of India is known from only 20 specimens and hasn't been seen since 1979. *Haploblepharus kistnasamyi* is a rare shark known only from three adult specimens, all collected from a small area near Durban, South Africa. Among skates, the Pita Skate (*Okamejei pita*) is also known from only a single female specimen collected in 1992 from the northernmost corner of the Persian/Arabian Gulf. The Ornate Sleeper Ray is known from two museum speci-

mens plus visual sightings by divers of four animals along 300 km of the eastern coastline of South Africa. The Java Stingaree (*Urolophus javanicus*) is another species known from a single specimen. It was collected near Java, Indonesia, 150 years ago and hasn't been found since.

When during the day are sharks most active?

Sharks as a general rule are more active at night than by day, although their opportunistic nature means they will take advantage of an easy meal, such as a dead fish or baited hook, almost anytime. A common pattern, then, is resting by day—alone or in groups, on the bottom or swimming slowly—then moving at dusk to a nighttime feeding area (see "Are sharks social?" in chapter 4). In many species, activity times are determined as much by food availability as by light levels. Plankton-eating sharks such as Whale and Basking sharks feed during daylight or we wouldn't have so many great photos of them. But their activity patterns as revealed by depth-recording tags suggest they also feed at night, following moving prey up and down in the water.

Manta rays—also plankton eaters—appear to feed both during the day and the night; some of the most dramatic videos of feeding mantas have been filmed at night. Many stingrays and skates forage actively during the day, but torpedo rays and angel sharks are more active at night. One exception is the Ornate Sleeper Ray, which vacuums up worms by day. White Sharks can feed anytime, but they tend to hunt seals during daylight and especially at dawn and dusk because that is when seals leave and return to their rookeries and when White Sharks can take advantage of backlighting to strike upward at a seal at the surface preparing for or recovering from a foraging dive.

How do sharks survive the winter?

Because they are oceanic animals and live mostly away from the shoreline, winter survival isn't a big issue for most sharks, especially tropical species. Some species that live at higher latitudes move seasonally but mainly to follow their food, migrating to places where food will be more abundant (see "Do Sharks Migrate?" above). Basking Sharks leave northern regions in the fall, moving south to warmer areas, but they then swim in cool depths, so they aren't gaining any great temperature advantage. Spiny Dogfish spend spring and summer off the west coast of Canada and move south in fall and winter.

Like many sharks and bony fishes, some rays and skates in the Northern Hemisphere move offshore or south to warmer waters in autumn, sug-

gesting avoidance of colder winter conditions. Atlantic Stingrays leave the Chesapeake Bay during the fall when surface waters cool and head south or offshore to warmer water. Cownose Rays move up and down the Atlantic coast of the United States, going north in the late spring and south in late fall. Clearnose and Winter skates move inshore in the spring and offshore in autumn.

How do sharks survive droughts?

Many bony fishes are "intertidal," meaning they live in tide pools or along shorelines that dry up once or twice daily at low tide. These fishes have many behavioral and physiological adaptations to help them through these relatively dry spells. Few sharks live in water so shallow that it goes dry on a daily or seasonal basis. An exception is the Epaulette Shark of Australia and New Guinea, which has the amazing ability to survive low oxygen conditions by lowering its blood pressure and heart and breathing rates, and switching off non-essential brain functions. Under laboratory conditions, Epaulette Sharks have survived up to an hour without oxygen at 30°C (86°F). (Recall that warmer water holds less oxygen at the same time that metabolic rates usually increase with increasing temperature.) This is apparently an adaptation for hunting in tidal flats at night when oxygen levels fall.

Some bony fishes such as lungfishes and Salamanderfishes dig into the mud when ponds lose their water in the dry season. We don't know of any freshwater sharks or rays that do this (on freshwater species, see "How many shark species live in rivers and lakes?" above). However, some sharks do live in places where rivers dry up seasonally, as in the Northern Territory of Australia. There have been instances in this region of Bull Sharks becoming stranded in pools as the former river dried up. An Australian fish biologist almost fell victim to a Bull Shark when he went to cool off in such a pool after a hot day of fish collecting. This fairly large Bull Shark had survived by eating animals that came to the pool to drink; dismembered kangaroo and camel skeletons on the shoreline were a clue that told him to exercise caution.

Are there sharks in the desert?

Deserts by definition are short on water, especially salt water. Where water occurs, it usually emerges from the ground at springs, which besides being too fresh for sharks also lack a necessary connection to the ocean that would allow invasion by sharks. So desert sharks are unlikely (although the movie *Sand Sharks* might lead you to think otherwise).

Do sharks have any enemies other than humans?

Sharks' greatest enemies, after humans, are other sharks. Several of the larger species—Tiger, Bull, White, and sleeper sharks—are known sharkivores. Cannibalism within a species is also common, as shown by Lemon Sharks in the Bahamas. One species of shark may even feed preferentially on another, as is true with the Broadnose Sevengill Shark (*Notorynchus cepedianus*), whose main prey off Tasmania, Australia, is the Gummy Shark. Some seemingly lethargic sharks are also successful predators on sharks; Tasselled Wobbegongs (*Eucrossorhinus dasypogon*), for instance, are able to ambush Brown-banded Bamboo Sharks on the Great Barrier Reef of Australia.

The shark-eat-shark nature of the ocean is probably the reason that females of many species go to particular nearshore "pupping grounds" to give birth to their young. These shallow inshore areas are infrequently visited by the males of the pupping species or by most large, predatory oceanic species.

Rays are another favorite prey of several shark species. Some hammerheads appear to be stingray specialists. Individual Great Hammerheads have been found with as many as 96 stingray barbs embedded in their mouth, throat, and tongue. Divers in the Bahamas watched a 3-m (10-ft) Great Hammerhead knock a Southern Stingray down with its flattened head, pin the ray to the bottom, then rotate on the ray and take bites out of the ray's wing.

Although we don't have such spectacular firsthand reports of sharks feeding on other rays, we do have circumstantial evidence. Manta rays observed at Tofo, Mozambique, have huge scars on their backs that indicate they were chomped on by large sharks and survived. Injured mantas, with fresh wounds, hover at "cleaning stations" on the Tofo reef, where various cleaner fish nibble at the wound to remove dead tissue and presumably

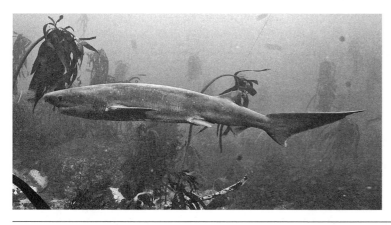

A Broadnose Sevengill Shark cruises a kelp bed off South Africa. Sevengills occur in temperate Pacific and Atlantic shallow-water locales and feed on other sharks, rays, and chimaeras. Photo by Austin Gallagher, http://austingallagher.com; used with permission

A Great Blue Heron captures a small Atlantic Stingray in shallow water in the Florida Keys. Moments later, the heron swallowed the ray whole. Photo by Kaye DeHays; used with permission

prevent infection (see "Do sharks socialize with other kinds of animals?" in chapter 4). Torpedo rays show up in the stomach contents of Broadnose Sevengills and Sixgill (*Hexanchus* spp.) sharks. Some torpedos have bite patterns on their backs indicating unsuccessful attacks by Sixgills, possibly because the torpedo was able to discharge its electricity before being swallowed by the shark. Despite possessing toxic barbs, young stingrays fall prey to other predators: Great Blue Herons have been observed spearing and swallowing small Atlantic Stingrays in Florida and Mississippi.

The prominent fin spines of chimaeras and Spiny Dogfish may offer some protection from predation. But chimaeras occasionally show up in the stomach contents of other ratfish as well as some sharks and seals. Even adult chimaeras are victimized, as shown by the presence of adult Australian Ghost Sharks in the stomachs of New Zealand Carpet Sharks (*Cephaloscyllium isabellum*) and School Sharks. White and Sevengill sharks happily munch on Spiny Dogfish.

A number of mammals kill and eat sharks of all sizes. Smaller shark species such as Horn Sharks are the prey of northern elephant seals (*Mirounga angustirostris*), the seals feasting on adults, juveniles, and egg cases. Juvenile Cape fur seals in South Africa have been seen capturing Puffadder Shysharks and tossing them repeatedly in the air, slowly dismembering them. The main object of this behavior doesn't appear to be feeding so much as playing with the shark, but the outcome is still usually fatal for the shark. Even if the seal doesn't eat the shark but drops it, Black-backed Kelp Gulls are always around to finish the shark off. Adult Cape fur seals have been filmed capturing and eating small Blue Sharks, also in South Africa.

Some of the biggest sharks fall victim to some of the ocean's biggest mammals. Sperm whales are best known for feeding on giant squid, but

whaling records indicate they also eat small deep-sea sharks. Sperm whales were also seen attacking a Megamouth Shark off North Sulawesi, Indonesia, the only known record of anything other than a Cookiecutter Shark attacking a Megamouth.

On an even larger scale, whale researchers have recently discovered that orcas (killer whales) feed on sharks around Vancouver Island, British Columbia, and farther north near Alaska. Packs of orcas gang up on sleeper sharks (probably Pacific Sleeper Sharks) and tear them apart. One feeding frenzy produced a large slick of floating shark meat and liver; the identity of the shark was confirmed by DNA analysis. Other observations suggest that orcas attack Blue Sharks farther south. Orcas have also been seen eating White Sharks off the California coast, with one often-seen female orca feasting on White Sharks at the Farallon Islands, offshore from San Francisco. Orcas have also been filmed attacking Whale Sharks, Common Threshers, Shortfin Makos, mantas, and stingrays.

Incidents in public aquariums suggest that sharks have other enemies, including some without backbones. Spiny Dogfish were disappearing mysteriously from a large exhibit at the Seattle Aquarium. Workers at the facility stayed late one night to see what was happening. They saw (and videotaped) a large giant Pacific octopus capturing and consuming a Spiny Dogfish. Whether such predation occurs in nature, and if so, how often, is unknown, but octopuses (not "octopi") are intelligent animals capable of capturing and eating a variety of prey, so it would not be surprising if the Seattle octopus was showing its natural predatory behavior.

Not all attacks on sharks are for the purposes of feeding. Dolphins and porpoises are reputed to ram sharks, apparently in defense of their young, although actual observations of this are hard to find. An incident at the Miami Seaquarium in which bottlenose dolphins attacked and killed a Sandbar Shark during the birth of a baby dolphin may have inspired the "conventional wisdom" that such events are common. In Hemingway's *The Old Man and the Sea*, the fisherman Santiago wonders how many sharks his marlin had killed when it was alive. What Hemingway based this idea on is unknown. A marlin being attacked by false killer whales in Hawaii speared a diver who tried to film the event, and fishermen off Nova Scotia and New York have observed Broadbill Swordfish turning on attacking sharks, so it isn't beyond reason to think that billfish may attack sharks, at least as a defensive maneuver.

Better documentation exists of other fishes ramming sharks. For reasons that remain mysterious, Giant Trevally (*Caranx ignobilis*), a relative of pompanos and jacks, attack and kill reef sharks much larger than themselves but don't eat them. Observations from the tropical Pacific (Saipan, Mariana Islands; Palau, Western Caroline Islands; Marshall Islands) have

documented large trevally singly or in pairs repeatedly head-butting reef species such as Blacktip Sharks and even Tiger Sharks. The trevally focus their ramming attacks on the sides and gill region of the shark. The shark often defends itself by trying—unsuccessfully—to bite its more agile attacker. Sometimes, a shark will attempt to flee to deeper water but be forced back into the shallows by the attacking trevally. The trevally ignore their own injuries, such as multiple cuts on their heads where they contact the sharks, and keep ramming. After repeated high-speed rammings, the shark, bleeding from its gills, sinks to the bottom and dies. Dissections of victimized sharks indicate damage to several internal organs that were probably the result of the rammings. Interestingly, trevally may ram more than just sharks. A Hawaiian spear fisherman was rammed by a large Giant Trevally, suffering three broken ribs.

How do sharks avoid predators?

Sharks avoid or deter predators in four major ways: (1) by running away; (2) by hiding, either in a hole or in plain view but with camouflage coloration; (3) by defending themselves with teeth, spines, toxins, or other structures and chemicals; or (4) by being hard to chew or swallow. Chondrichthyan fishes do all four.

By Running Away. Running away can mean just trying to outswim your predator, or it can mean moving to a safer locale. It all depends on whether you are faster or more maneuverable than your pursuer. Juvenile Lemon Sharks in the Bahamas take advantage of an incoming tide to move into shallow areas around mangroves that larger, faster Lemon Sharks—one of their main predators—avoid. Young Largetooth Sawfishes spend their early lives in similar habitat.

Knowing when to run away is obviously important because getting a head start can make all the difference. Some stingrays improve their chances of detecting a predator and fleeing by associating with other stingray species that are better at keeping a watchful eye for predators. Cowtail Stingrays (*Pastinachus sephen*) in western Australia rest during the day on sandy bottoms. When doing so, they prefer to rest near another species of dasyatid ray, the Reticulate Whipray, rather than with other Cowtail Stingrays. Reticulate Whiprays have a quicker flight response than Cowtails when a predator or a boat with researchers approaches.

It's not just small sharks that run away from predators. Killer whales are known to attack and eat White Sharks. Researchers studying White Sharks at the Farallon Islands off California noticed that their study animals became scarce after one shark was eaten by a killer whale. One tagged

shark immediately left the area and went to Hawaii. How other sharks were aware of the attack is something of a mystery, although the commotion and splashing during the attack and the copious amount of White Shark blood and body fluids in the water could have been detected by other sharks.

By Hiding. Many smaller sharks hide from predators. Some small night-active sharks spend their daytime resting in caves and crevices, often in the company of other members of their species (see "Are sharks social?" in chapter 4). Other sharks, as well as skates and rays, rely on blotchy, camouflage coloration to make themselves hard to see when they are lying motionless on the bottom (see "Are any sharks colorful?" in chapter 3). Wobbegongs also have tassles and skin growths that help them blend in better with the variable colors and structures of an algae-covered reef. On sandy bottoms, angel sharks, skates, and stingrays cover themselves with sand with just their eyes and maybe their spiracles showing, allowing them to see and breathe while being hard to be seen.

By Defending Themselves. Defending yourself with teeth and spines is effective, although it means you actually have to battle a predator that is in all likelihood bigger than you. All but the biggest sharks (Whale, Basking, Megamouth) have sharp teeth and strong jaws that can deliver an injury-causing bite. Some species, such as Nurse Sharks, are even flexible enough to stick their tails in their mouths, which makes pulling the tail of a Nurse Shark lying on the bottom a very bad idea. Torpedo rays have the unusual option of defending themselves with their electric shocking equipment.

Many prey species that have good defensive mechanisms advertise the fact to predators. This makes sense because it's better to not be attacked or swallowed, if at all possible. Several stingrays, such as Crossback Stingarees (*Urolophus cruciatus*), Southern Stingrays, and Shorttail Stingrays raise their tails and stings above their heads, scorpionlike, when approached by a diver. We assume that they do the same when approached by a potential predator. Some torpedo rays advertise their defensive armament by arching their back and tumbling and twisting in the water, a behavior that may be a warning display, or may just present the strongest discharge surface to an attacker. The most colorful electric ray, the Ornate Sleeper Ray of South Africa, arches its back and raises its tail in what might be a warning to a predator that it can defend itself with an electric shock.

Skates aren't as well-equipped as torpedo or sting rays, and their defenses appear to be more passive. A few divers have seen skates roll themselves up into balls, curling their pectoral fins down to protect their more vulnerable bellies and exposing their backs, which are often adorned with

hard thorns. Southern Thorny Skates (*Amblyraja doellojuradoi*) do this when caught in nets.

Divers in Norway have made intriguing observations of the defensive behavior of skates. Divers who bump into or prod juvenile Thorny Skates have been surprised to see the skates quickly roll up hedgehog-style with their pectoral fins curled around their bellies, essentially lying on their backs with their heads up and their thorny tails presented to the intruder. What makes this behavior intriguing is that the surface presented to a would-be predator is the part where the electrogenic tissue of skates is concentrated. One of the nagging questions in elasmobranch anatomy and behavior is the function of this tissue in skates. Skates emit a relatively weak electric charge, nothing like that of a torpedo ray. The function(s) of skate electric discharge remains something of a mystery, although mating skates discharge electricity, implying some reproductive purpose. The major predators of Starry Skates are two species of sharks, the Greenland Shark and Tope. One idea proposed is that a skate that is discharging its electric organs might somehow jam or overwhelm the extreme electric sensitivity of a predatory shark. It's an idea worth testing.

Most chimaeras have first dorsal fin spines with a sawtooth back edge, covered with a venomous toxin. Researchers and fishers unlucky enough to get stuck by a ratfish report that it is a painful experience, the wound remaining red and swollen for several days. Presumably, natural predators have the same experience. The toxic spines of Spiny Dogfish may function similarly.

By Being Hard to Eat. Anything a prey animal can do to make itself harder to swallow works to its advantage. Tough skin and stiff spines, especially long spines with toxins, deliver a message of inedibility. Stiff spines also increase an animal's overall size, assuming the predator must swallow its prey whole (this works better as a defense against predatory bony fish that can't chop up their prey the way that most predatory sharks can). Bramble sharks (*Echinorhinus*, Echinorhinidae) live as deep as 900 m (2,970 ft). Their unique large thornlike denticles might give them physical protection from deepwater predators such as somniosid sleeper sharks and sperm whales. The tough skin and spines of Horn Sharks apparently work to deter predators. A Pacific Angelshark (*Squatina californica*) was filmed inhaling a juvenile Horn Shark, only to spit it out shortly afterward, probably because of the boxlike body and stiff spines that give the shark its name.

The idea that making yourself bigger prevents a predator from swallowing you underlies the anatomy and behavior of several shark species. Puffadder shysharks (*Haploblepharus* spp.) and Pajama Catsharks (*Poroderma africanum*) curl into a circle and cover their eyes with their tail when threat-

ened; this is why they are known as shysharks or even doughnut sharks. This behavior may make these small (60-cm, 24-in), slender sharks bigger and therefore harder to swallow.

Swellsharks (*Cephaloscyllium*) go a step further in being able to increase their size when threatened. Swellsharks are unique among sharks because they can swell their entire body by filling their stomachs with water, a behavior better known in bony fishes such as pufferfishes. When threatened, a swellshark bends its body into a U shape, grabs its tail fin in its mouth, and swallows water, swelling its body to twice its normal size. This makes the shark harder to grab, and if the shark happens to be tucked away in a crevice, it would also make it difficult to extract. So, inflatable sharks are more than pool toys.

Do sharks get sick?

Very few captured sharks are found to be diseased. Sharks have strong, efficient immune systems. Their immune reaction to disease may be faster than it is in other vertebrate groups such as mammals. In sharks, immune cells are produced in the spleen, thymus, and other places, but not in bone marrow as they are in mammals (remember: sharks don't have bones). Immune cells are already in the shark's blood and can multiply quickly in response to a challenge from a disease-causing substance or organism. But this doesn't mean sharks are disease-free. They suffer from a variety of ailments and illnesses, including bacterial infections, liver disease, meningitis, tumors, and the effects of a huge variety of internal and external parasites. Parasites are discussed in the next question.

It was long thought that sharks were immune to cancer. As a result, sharks were killed to extract potential cancer-curing remedies. Shark cartilage was promoted as an anticancer drug because laboratory studies showed that cartilage extracts could slow the growth of blood vessels that often occurs when cancer cells proliferate. Shark cartilage powder has sold for as much as $145 a gram ($65,685 a pound), and at one time a single plant in Costa Rica processed 235,000 sharks per month to make cartilage pills. But sharks do get several kinds of cancer, including cartilage cancers. Also, real trials with real people showed no beneficial effect from cartilage pills. Regardless, sharks are still processed for their cartilage, and cartilage capsules can still be purchased via the Internet to treat cancer and joint problems (see "Are any medicines made from sharks?" in chapter 8).

Do sharks have parasites?

We like to think of big sharks as residing at the top of the food pyramid, surpassed only by an occasional killer or sperm whale. But in reality, each shark is a tasty, swimming platform for a zooful of smaller parasitic animals that feast on its blood and body tissues. Shark parasitology is a popular field of study because of the intimate evolved responses between a vast diversity of parasites and their shark hosts. Some parasites live in or on only one type of shark and attach only in a particular place inside or outside the shark, and only during a particular part of their life cycle. (Earlier life stages first parasitize invertebrates, and then the small fishes that eat the invertebrates, linking parasite life cycles to the entire food web.) About 1,500 different parasites are known from sharks, skates, and rays, with each elasmobranch species host to between 3 and 14 unique species; no doubt, many more will be discovered.

Remarkably, according to shark parasitologists Janine Caira and Claire Healy, "no organ system of elasmobranchs has escaped the attention of one or more groups of parasites." If we performed a CT scan across a shark, from head to tail and including internal organs, we would find parasites in the following shark organs:

- Skin—One type of copepod crustacean (a relative of shrimp and crabs) lives only on the pelvic fins of Blue Sharks, and one monogenean flatworm lives only on the undersides of certain skates. Other skin parasites include a snail that lives on the dorsal side of torpedo rays, 20 species of leeches, 275 species of copepods, 200 other monogenean flatworms (8 of which occur on chimaeras), three types of nematode roundworms, barnacles that attach to fin spines and flesh, amphipods (another shrimp-like crustacean), branchiuran fish lice, isopods (same group as pill bugs), and a few cestode tapeworms and digenean flukes.
- Eyeballs—Eyeball parasites include several copepods (see below regarding *Ommatokoita* on Greenland Sharks), a leech on Spiny Dogfish, and a fluke on Whitetip Sharks.
- Olfactory bulbs (the complex foldings inside the nose that lead via the olfactory nerve to the brain)—Among nasal parasites are ostracod seed shrimps, isopods, copepods, roundworms, leeches, and flatworms. One copepod infects the Barbeled Houndshark (*Leptocharias smithii*), the head of the parasite being lodged in the olfactory lobe of the brain and the tail resting in the olfactory bulbs. A roundworm has also been found wrapped around the optic nerves and much of the brain of a Spotted Eagle Ray.

- Lateral line—A copepod lives inside the lateral line canals of the Porbeagle.
- Spiracles—Three families of copepods and some leeches live inside these small breathing holes on the top of the head of some elasmobranchs.
- Gills and gill chambers—Gills are infested with leeches, roundworms, ostracods, *Argulus* fish lice, isopods, digenean flukes, monogenean flatworms, an occasional tapeworm, and a host of copepods from 11 different families. In Blue Sharks, some copepod species live only on the walls between the gills, others on small gill filaments, and others in the channels of outflowing water; still others draw blood from specific blood vessels in the gills.
- Digestive system—The different regions of the digestive system, from mouth to rectum, each have their own catalogue of specific parasites. The mouth and esophagus are home to leeches (some of which occur in only one shark or ray species), copepods, flukes, flatworms, roundworms, and barnacles. The stomach, despite its hostile chemical nature, houses flatworms, roundworms, spiny-headed worms, isopods, and numerous tapeworms. The spiral valve intestine has its own group of parasites, including spiny-headed worms, flukes, flatworms, 50 species of roundworms, and several hundred tapeworm species, some of which live only in specific regions of the spiral valves. One family of tapeworms lives only in chimaeras. One genus of spiny-headed worms is specific to the potamotrygonid freshwater stingrays of South America. Some batoid skates and rays can house as many as 10 species of tapeworms in the spiral valve. The rectum and cloaca at the posterior end have their own list of flukes and flatworms, copepods, leeches, and isopods.
- Gall bladder, bile ducts, pancreatic ducts, liver, etc.—A few flukes, tapeworms, and nematodes occur only in these organs and tissues. One nematode also lives in the spleen of the Big Skate (*Raja microocellata*), and roundworms have been found in shark kidneys. The body muscles of sharks are relatively parasite free, except for an occasional roundworm, unlike the situation in bony fishes.
- Heart and blood vessels—The circulatory system is occupied by roundworms, flukes, flatworms, mites, isopods, and an occasional eel (see below).
- Reproductive tracts—Male gonads are relatively parasite free (an occasional tapeworm has been found); females house nematodes, tapeworms, flatworms, flukes, leeches, copepods, and isopods variously in their ovaries, oviducts, and uterus. Developing embryos are already parasitized externally by leeches and copepods.

Eyeball from a Greenland Shark infected with a parasitic *Ommatokoita elongata* copepod. The photo shows a female *Ommatokoita* anchored in the shark's eyeball via a pair of modified legs. Her egg cases are the white bulb-like structures protruding from the eyeball to the right. An infected shark may be essentially blinded by the copepod but is still able to feed successfully. Photo courtesy of George Benz

Most internal parasites are small, wormlike animals. But others are large and have backbones. For example, when researchers cut open a 395-kg (870-lb) Shortfin Mako Shark caught off New York, they found two live Snubnose Parasitic Eels (*Simenchelys parasiticus*) curled up happily inside the shark's heart. The eels, which are true bony fishes, were about 20 cm (8 in) long, took up most of the space in the heart, and had shark blood in their guts. It was difficult to say how long they had lived there, but scar tissue in the heart indicated that the eels had been there quite a while.

Most of the external parasites on sharks can move across the skin and around the scales. They may find the hard dermal denticles and thick skin of sharks a challenge for attachment. But others seek smoother, softer body parts where they can anchor themselves. One example is a particular type of copepod, *Ommatokoita elongata*, which targets the eyeballs of Greenland Sharks. *Ommatokoita* is a good example of a habitat-specific parasite; it is a 5-cm (2-in) copepod that penetrates the shark's eyeballs and feeds on the surface of the cornea. The parasite isn't fatal; sharks with parasites embedded in both eyes are functionally blind but still grow to large sizes and feed successfully.

Reflecting on all this information about internal and external parasites can change how we look at sharks. When you see a shark swim by in the wild or at a public aquarium, try to realize that what you are seeing is not just a single animal but more like an entire, highly evolved, mobile ecosystem.

Sharks: The Animal Answer Guide

How can you tell if a shark is sick?

A healthy shark swims normally. A sick shark would appear disoriented, swimming erratically or lying on the bottom on its side or back. Such an animal would be eaten quickly by other sharks or by a variety of scavengers in the wild. Public aquariums have specially trained veterinarians to deal with sick animals, but there is little the average home aquarium keeper can do. This is one of the reasons we advise against keeping sharks in fish tanks at home.

Are sharks good for the environment?

In some areas, rays serve as "ecosystem engineers" by digging up clams, crushing their shells and turning them into sand, and moving sediment around in the process. Their digging often pushes soft sediment aside, exposing hard bottom and objects that become settling habitat for a variety of invertebrates. A study in western Australia estimated that three ray species reworked 42% of the sediment in a bay each year.

Rays are an example of the different roles played by different species in natural ecosystems. When species are eliminated, relationships and processes are affected. Predators play important ecological and evolutionary roles, preventing overpopulation by reducing prey numbers and removing weak or sick individuals. Predators therefore maintain healthier prey populations with less disease and fewer genetic defects. When predators are removed, prey populations become overpopulated. Deer populations throughout the United States have reached unnaturally high numbers and have become pests because their natural predators (wolves, mountain lions) have been "controlled." Deer then outstrip their food resources and starve, contract fatal epidemic diseases, and cross highways too often. Rabbit populations skyrocket and strip vegetation bare in the absence of predators.

The same rules hold for marine systems. Coral reefs that are overfished become covered with coral-killing algae because the fishes that ate the algae have been removed. When sea otters were hunted almost to extinction in the North Pacific, the sea urchins that they used to eat increased. Urchins eat algae such as kelp and consequently create "urchin barrens," which are rocky areas devoid of plants. Kelp beds provide habitat for a vast array of nearshore fishes and invertebrates. Without kelp, these animals are homeless, and their populations diminish.

Sharks are at the top of the food chain in many systems (ignoring their parasites for the moment). When they are overfished, their absence affects the different levels of the food chain below them.

Chapter 6

Reproduction and Development

How do sharks reproduce?

We know a great deal about certain aspects of elasmobranch reproduction. Or at least we know much about the things that can be studied by dissecting dead fish. We know comparatively less about the behavior and ecology of reproduction, although our understanding is growing.

Certain behaviors occur in several of the species whose reproduction we have been able to observe. Mating in elasmobranchs can be an elaborate and sometimes violent affair, at least by human standards. For example, males follow closely behind females for varying lengths of time—a few minutes in Nurse Sharks but as much as six hours in Basking Sharks. One or several males may follow a female, depending on species. A male may swim with his snout almost in contact with the female's tail or even her cloaca. In manta rays, males follow closely behind a female, duplicating her every swimming movement. As many as 25 male mantas will follow a single female in what is called a "mating train," making for a group ballet performance. Mating trains may last for more than a day. Courtship displays also occur. For instance, a male Scalloped Hammerheads will approach a female that is part of a group, bend his body, and stick out a modified pelvic fin clasper. If the female is receptive to this display, the couple swims off as a pair and copulates.

Actual mating begins when a male grabs onto a female the only way he can, with his teeth. (In manta rays these are tiny, are only on the lower jaw, and probably function just for mating.) A male may bite down on a female's pectoral fin, gill area, or body, holding her while he inserts one of his claspers into the female's cloaca. Internal contact is maintained by barbs, hooks, claws, and spines at the end of the clasper. In Blue Sharks, the clasper ends

Few sharks have been observed mating, and even fewer have been filmed. This photo was taken just as a male Nurse Shark grasped the right pectoral fin of a female. The female is to the right in the picture, on her left side. Photo © Jeffrey C. Carrier, all rights reserved; used with permission

in an umbrella-shaped structure. In smaller, flexible sharks, the male may wrap himself around the female's body, whereas larger, stiffer sharks swim side-by-side or belly to belly. Male Spotted Eagle Rays grab a pectoral fin, rotate, and swim upside down under the female while inserting a clasper. Actual copulation may last less than two minutes in Eagle Rays or up to four hours in Clearnose Skates (*Raja eglanteria*).

Regardless of species, once a clasper is inserted, the male pumps sperm into the female's reproductive tract, the sperm propelled in part by sea water pushed from a muscular bladder at the base of his claspers. Females often bear bite marks and ragged scars on parts of their body from these mating encounters, some the result of a female pulling herself away before a male can insert a clasper.

Thanks to a multiyear study conducted by researchers in the Dry Tortugas marine reserve in south Florida, the shark species whose mating behavior we know best is the Nurse Shark. Nurse Sharks are common around southern Florida, live in shallow water, and can be observed while they are mating if researchers approach them carefully. Sometimes several males pursue a female simultaneously, until one grabs her by her pectoral fin and attempts to carry her into deeper water where he has room to copulate. A female may avoid the advances of a male by moving into shallower water and even burying her pectoral fins in the sand, making it harder for a male to grab her. When a male does succeed in grasping her fin, she may then quickly roll over on her back above the male, trying to twist loose, and push her ventral surface and cloaca in the air, waiting for him to give up. (Some female rays avoid copulation by burying in the sand, slapping their pectoral fins on the surface, and sometimes stabbing a male with their barb.) She may shake loose but in the process sustain damage to her pectoral fin

Reproduction and Development

from the male's teeth, with the result that the fins of many Nurse Shark females bear obvious tears and scars. Because of females' resistance, fewer than 10% of mating attempts actually end in successful clasper insertion. A female Nurse Shark may mate with several males during June, which is the primary mating season in south Florida.

The one aspect of reproduction that all chondrichthyans have in common is internal fertilization of the female's eggs. Internal fertilization and live birth have apparently been characteristics of sharks since their earliest evolution. Fossil sharks have structures that look much like the claspers of modern sharks, and some fossils have shark embryos inside apparent mothers. Violent mating behavior may seem a strange way to reproduce, but it too has been around for much of shark evolution, and adaptations to it show up in several aspects of elasmobranch anatomy. In Blue Sharks, a female's skin is thicker than the male's teeth are long, whereas males have thinner skin. The teeth of some male scyliorhinid catsharks (*Halaelurus*, *Apristurus*) are longer and have fewer side cusps than those of females, presumably to aid in holding onto females. Outside of the mating season, male Atlantic Stingrays have blunt, rounded teeth similar to those of females, an appropriate design for crushing clams. As mating time approaches, mature males develop teeth with long, sharp edges that are better for gripping females.

A male may court and mate successfully, but that doesn't necessarily mean he will fertilize eggs, or at least not immediately. Females can store sperm, probably in the shell gland, for periods of a few weeks to more than a year, depending on species. Sperm storage thus delays the time when eggs are actually fertilized. For example, female Blue Sharks mate before they are sexually mature. They store sperm, not releasing it until 10 to 20 months later. Other sharks known to store sperm include Whitespotted Bamboo Sharks (*Chiloscyllium plagiosum*) and Tiger Sharks, both for three to six months; Nervous Sharks (*Carcharhinus cautus*, one month); Topes (five months); Whiskery Sharks (*Furgaleus macki*, six months); and Chain Catsharks (*Scyliorhinus retifer*, one year). Lamniforms are not thought to store sperm.

Courtship and mating have been observed in a few skate and ray species. Males follow females and maintain contact by biting, as with the Eagle Rays mentioned earlier. In some skates, sexually mature males have additional structures that either help them hold onto a female or perhaps stimulate her to mate. These structures are thorny, hooklike spines around the top of their heads or on the upper surface of their pectoral fins.

Skates and chimaeras are egg-layers with internal fertilization. Male chimaeras have a special stalked, hooklike structure on their heads that ends in a spiky bulb. Instead of biting down on a female with his parrot-like beak, a male uses this cephalic (head) clasper to hold the female's pectoral

Sharks: The Animal Answer Guide

The skin of female Blue Sharks (*right*) is thicker than the skin of males (*left*), probably to prevent muscle damage during courtship, when males bite females. (The ruler length is 15 cm, or 6 in.) Photo by H. L. Pratt, from H. L. Pratt, Jr., "Reproduction in the Blue Shark, *Prionace glauca*," U.S. Fishery Bulletin 77 (1979): 445–70

fin against his forehead while he inserts a multi-lobed pelvic fin clasper into the female.

We know very little about the cues that male elasmobranchs use to determine if a female is ready to mate. Chemical attractants produced by females are a likely candidate, given how sensitive sharks are to odors and the way that males follow closely behind females. It's also possible that the electric sense comes into play, at least in batoids like the Round Stingray. Round Stingray males can locate buried females by detecting the electric field produced by a female. Skates in particular seem to use their electricity in male-female interactions. Some female skates, such as Little Skates, emit weak electric pulses at a frequency of 2 to 3 Hz, which is the frequency to which the males are most sensitive.

Do sharks lay eggs or do they give birth to live young?

Both. Once fertilization happens, young may be deposited (laid) outside, protected by a leathery, shell-like structure. Or they may develop in the mother, either inside a shell-like structure for a while or exposed the entire time while developing in the mother's uterus. Most sharks and all rays give birth to live young, whereas rajid skates and chimaeras lay eggs. Among the few egg-laying selachian sharks are heterodontid horn sharks, scyliorhinid cat sharks, hemiscyliid bamboo sharks, and the stegostomatid Zebra Shark. All these egg-layers are small (less than 1 m, or 3 ft, long) and swim near or lie on the bottom to begin with. They lay their eggs on the bottom or attach them to vegetation or other objects. A few shark groups that lie on the bottom, including orectolobid carpet sharks, give birth to live young, with eggs developing inside the mother first.

Live-bearing sharks and rays fall into one of two groups depending on how the embryo gets its nutrition before birth. In the more common group, the young use up a yolk supply to which they are attached from the start, with no additional nutrition provided by the mother. In the other group, the developing embryo receives nutrients other than yolk from the mother.

All live-bearing sharks except some of the carcharhiniform requiem sharks and the lamniforms (White Shark and relatives) use yolk sack nutrition. Some female squaliform dogfish and triakid houndsharks supplement yolk nutrition by secreting mucous from their uterus walls that is ingested by the embryo. In some stingrays, the mother secretes fatty substances, called uterine "milk," that the embryo eats. Another type of mother-provided nutrition is called oophagy, literally "egg-eating." Here, the mother keeps on producing unfertilized eggs, which are eaten by the developing pups. Oophagy occurs in lamniforms (White, makos, and others) and pseudo-triakid false cat sharks. One lamniform, the Sand Tiger, carries oophagy one step farther: the first embryo to develop eats not only eggs but also its later-developing siblings. Sand Tigers give birth to only two 1-m-long (3-ft) young, one from each uterus. The surviving embryos already have well-developed, functional recurved teeth in both jaws when only 45 mm (1.5 in) long. Cownose Ray pups also eat unfertilized eggs and their siblings, the result being that a female gives birth to only one pup a year.

One group of live-bearers, composed of five families of carcharhiniforms, has evolved a direct connection between mother and pup, similar to the placenta and umbilical cord that connects a developing mammal (dog, giraffe, human) to its mother. The umbilical cord in these sharks develops from the tube that connected the developing shark to its yolk sack. The tube has many small outgrowths and is attached to the uterine wall; nutrition passes from mother to young along the tube's outgrowths once the nutrition in the yolk sack is used up. Young sharks in this group are born with a placental scar on their chests (so, yes, many baby sharks have a belly button).

The pups produced by live-bearing sharks and rays are relatively large replicas of their parents, able to feed on their own and defend themselves from predators. However, large size and good defenses create problems during the birth process. These problems are solved by special adaptations. Hammerhead shark pups would have a difficult time getting their expanded heads through the birth canal, so the lobes of the hammer are soft and pliable at birth and then stiffen soon after. Squalid dogfish sharks have stiff, sharp spines in front of their dorsal fins. The spines of embryonic dogfish are covered with pads of tissue until after the young are born. Mother sawfish are similarly protected from their developing and emerging young. The saws of fetal sawfish are at first soft and enclosed in a rubbery envelope. When Bat Rays are born, the pectoral wings are folded over the body

Sharks: The Animal Answer Guide

Placental connections in sharks. A newborn Atlantic Sharpnose Shark, showing the umbilical cord with the outgrowths that take up nutrients from the mother. The cord ends in a placenta that attaches to the uterine wall of the mother. Photo by W. C. Hamlett; used with permission

to facilitate passage, and the venomous barb is flexible and covered in a sheath that is lost shortly after birth.

In egg-laying elasmobranchs, the developing embryo gets its nutrition from the yolk that was part of the unfertilized egg. Aside from some of the carcharhinid reef sharks, chondrichthyan eggs are not round like bird eggs but are more often rectangular and flattened. The corners of the egg shell often have horns and sticky, stringy extensions ("tendrils") that help anchor it to a structure. The usual situation is one embryo per egg case, although in two skates (Big Skate and Mottled Skate, *Raja pulchra*), two embryos occur frequently and as many as seven embryos have been found in a single egg case. Chondrichthyan eggs range greatly in size and shape, from the more common flat rectangles to complex, corkscrew-shaped cases in heterodontid bull sharks. Skate egg cases can be as small as 4.6 cm long by 2.7 cm wide (1.8 by 1 in) in the Little Skate, to as large as 31 by 19 cm (12 by 7 in) in the Big Skate. The Blue Skate of the western Atlantic is the largest skate in the world, but its egg capsules are no more than 24 cm long by 14 cm wide (10 by 5.5 in). The empty egg cases of skates often wash up on beaches and are referred to as "mermaid purses" by beachcombers. Egg cases differ in size and structure depending on species, often allowing identification of the skate that produced it (see www.eggcase.org).

The developing embryo inside the egg is protected by a leathery or horny shell that was produced by a shell gland in the mother's reproductive tract. As the embryo grows and uses up its yolk supply, its metabolism increases. Oxygen demand goes up, and waste products accumulate. To

Reproduction and Development 117

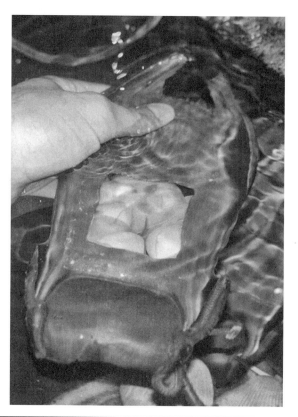

Egg case, or mermaid's purse, from a Big Skate, *Raja binoculata*. This North Pacific species is the largest skate in North America, reaching a length of up to 2.4 m (7.9 ft) and a weight of over 90 kg (200 lb). A section of the shell has been cut away to reveal several developing embryos inside the case. Some large skates are overfished because they are born at a size that is already too large to escape through the mesh of fishing nets. Photo by Gene Helfman

renew the water supply inside the egg, embryonic skates lash their tails back and forth, moving water in and out of the case via small slits in its side. This movement in the semitransparent egg case and the resulting water currents can attract predators such as other skates as well as sharks, bony fishes, some snails, and marine mammals. If something bumps into an egg case, or if the egg case is exposed to a weak electric discharge such as the ones produced by other skates, the embryo stops moving its tail.

Chimaeras are also egg layers. A female produces relatively large eggs (15–20 cm long by 5–7 cm wide, or 6–8 in by 2–3.5 in) surrounded by a horny shell, one embryo per egg case. Chimaera egg cases are rounder and more elongate than skate and shark egg cases. Eggs are laid on the bottom, usually two at a time, one from each oviduct. Several pairs of eggs may be laid over the course of a reproductive season, with egg-laying bouts occurring every one or two weeks. Small slits in the posterior end of the egg case allow water to enter and metabolic wastes to be carried away.

How long do female sharks hold young in their body?

Gestation is the period from egg fertilization to birth. It's another way of discussing embryo development. The human gestation period is nine

Sharks: The Animal Answer Guide

months. Elasmobranchs as a group have relatively long gestation periods, an important factor in understanding their conservation because a slow reproduction rate means that when a population is overfished, individuals are replaced slowly.

Gestation periods vary greatly among sharks and rays. Some of the shortest periods for live-bearing elasmobranchs are found in rays such as the Atlantic Stingray, which carries its young for only 3 or 4 months. Manta rays, in contrast, have a 12-month gestation period. Nurse Sharks have a relatively short period of about 5 months. Pacific Angelshark young gestate for 9 to 10 months, which appears to be about average for the family. Blue Sharks take longer, 9 to 12 months, as do Blacktip, Finetooth, Atlantic Sharpnose (*Rhizoprionodon terraenovae*), and several other carcharhinids. Smooth Dogfish have an annual reproductive cycle. Sphyrnid hammerheads appear to also have an annual cycle, with young gestating for 8 to 11 months after fertilization.

Size isn't necessarily a good indicator of gestation period. Spiny Dogfish, a species that seldom exceeds a meter or so (i.e., 3–4 ft) in length, have a two-year gestation period. Basking Sharks are suspected of holding their young for at least that long, but we don't know enough about Basking Shark biology to be able to say for sure.

How long does it take for a baby shark to hatch from an egg case?

Hatching times also vary greatly among species. Data on hatching times from egg-bearers are more accurate than those from live-bearers because we can monitor the development of eggs that have been laid. It is much harder to know when a female live-bearer mated and, because of sperm storage, just when eggs were fertilized. Fortunately, captive animals will lay eggs in an aquarium, and the entire developmental process can be observed directly. Hemiscylliid bamboo sharks hatch relatively quickly, in 2 to 6 months. Heterodontid Horn Shark eggs hatch in 7 to 9 months; other heterodontids, in 5 months to a year. Hatching times in scyliorhinid catsharks and swellsharks vary greatly because some species lay eggs when the young are just beginning to develop whereas other species retain the eggs almost up to the time of hatching. Brown Catshark (*Apristurus brunneus*) eggs were collected near Monterey, California, and brought into the Monterey Bay Aquarium. Pups hatched 27 months later; data from other locales suggest at least one year to hatch in nature. Swellshark eggs take 7.5 to 10 months to hatch, depending on water temperature. The South African Lined Catshark (*Halaelurus lineatus*) lays eggs further along in development, the young hatching in only 3 to 4 weeks.

Skate incubation periods are highly dependent on temperature; babies develop more quickly at warmer temperatures. Thorny Skate incubation can take as little as a couple of months to as long as a year. Little Skate eggs may hatch in 6 to 9 months in Connecticut waters but may take a year to hatch farther north in Maine.

Hatching times for chimaeras are known for two species, the Australian Elephant Fish and the Spotted Ratfish. Hatching appears to take between 5 and 12 months, the shorter time applying to the Australian Elephant Fish. The baby chimaeras that emerge from these eggs are about 10 to 15 cm long (4–6 in).

Where do sharks lay their egg cases or give birth?

Many elasmobranchs leave their normal habitats to give birth or lay eggs. Females usually move into shallower water, often near shorelines. Bluntnose Sixgill Sharks (*Hexanchus griseus*) commonly occupy oceanic depths of 500 to 1,500 m (1,650–5,000 ft). Pregnant females move into shallow Puget Sound, Washington, and juveniles are fairly common at depths of less than 30 m (100 ft). Broadnose Sevengill Sharks are a coastal species that occurs to depths below 100 m, but females move into shallow areas of Humboldt and San Francisco bays to give birth. Lemon Sharks frequent shallow reef areas but move into even shallower mangrove forests to give birth. Scalloped Hammerheads are well known for migrating to and from oceanic seamounts near deep water; even so, they give birth in shallow regions of the Gulf of California and Kaneohe Bay, Oahu Island, Hawaii. Scalloped Hammerheads and Smooth Hammerheads (*S. zygaena*) move into the innermost sections of Ariake Bay in southern Japan in summer to give birth. This same shallow section of Ariake Bay is a nursery area for four shark and eight ray species.

Female Leopard Sharks and Whitetip Reef Sharks also move into shallow, warmer bays before giving birth. It has been suggested that by moving into areas with higher water temperatures, females increase the growth rate of their developing pups; the larger and presumably faster pups would then be better able to find food and avoid predators.

Some holocephalan chimaeras also move inshore to reproduce. Rabbitfish normally inhabit Atlantic waters at depths between 300 and 1,250 m (990–4,125 ft) but migrate in spring and summer to water less than 100 m (330 ft) deep to lay eggs. Australian Elephant Fish also migrate inshore to shallow water bays during their summer-to-autumn egg-laying season and then move back offshore in winter and spring.

Because most larger sharks live in deep water offshore, it is assumed that by giving birth in shallow nearshore areas, females are reducing pre-

dation on their newborn pups. However, not all predators are deterred by this tactic. For example, some sharks will follow female Spotted Eagle Rays during the birthing season and snatch up the ray pups shortly after they are born. As mentioned below, the birth of Broadnose Sevengill Sharks in San Francisco Bay appears to be timed to take advantage of the later appearance of young Brown Smoothhounds (*Mustelus henleyi*), on which the Sevengill young feed.

Do sharks give birth or lay their egg cases at the same times and places every year?

We now know from tagging and genetic sampling that, in some species at least, the same females return to the same shallow nursery areas multiple times. Additionally, the time for giving birth is predictable and short in many species. Nurse Sharks in South Florida have a late November to early December "pupping" season. Four different elasmobranchs pup in San Francisco Bay, spreading the pupping season across spring and summer, although each species has a relatively short season. Female Bat Rays arrive in late March and give birth two weeks later, in early April. Leopard Sharks congregate on eelgrass-covered mudflats in small bays and pup in late April to early May. Broadnose Sevengill Sharks release young between April and May. Brown Smoothhounds pup a little later, from mid-May to early June.

Such "traditional" birthing and nursery grounds are even suspected for some of the biggest species, such as Reef Mantas in the Indian Ocean. These giants aggregate in the Maldive Islands in October and November, where 80% of the mantas are females and many are very pregnant. Mating occurs shortly after the females give birth, although no one has yet seen a manta being born in the wild.

The available evidence on chimaeras also suggests traditional times and places for egg laying. For example, Australian Elephant Fish lay eggs predictably during summer and autumn in Port Phillip and Westernport bays, southern Australia. The eggs then hatch in September and October, the Australian springtime.

In species with fixed times and places, protection of such pupping areas is important to conserve the species involved.

How many young does a mother shark produce?

The litter size, or number of young produced, varies among species from one or two (manta rays; Sand Tiger; Lowfin Gulper Shark, *Centrophorus lusitanicus*) to 300 (Whale Shark, Pacific Sleeper Shark). Most chondrichthyans produce intermediate numbers, egg-laying species generally

Number of young produced by some live-bearing sharks

Species	No. of pups per litter
Threshers, Longfin Mako, pseudotriakid gollumsharks, several centrophorid gulper sharks	2
Carcharhinids, except for Tiger and Blue sharks	3–10
Spiny Dogfish	6–7 every 2 years
Basking Shark	6
Frilled Shark	6–12 every 1–2 years
Many triakid houndsharks	5–15
Pristiophorid saw sharks	6–19
Sharpnose Sevengill Sharks (*Heptranchius perlo*)	6–20 annually
Sawfishes	8–20 pups every other year
Scalloped Hammerhead	1–41 pups every other year (av. = 16)
Squalid sharks	Up to 32
Nurse Shark	25–50
Tope	6–50 (increasing with size of mother)
Hexanchid cow sharks	20–100
Prickly Shark	114
Tiger and Blue sharks	70–135
Whale and Pacific Sleeper sharks	300
Guitarfishes and stingrays	6 per year

producing more than live-bearing species. In many species, litter size increases as a female grows because larger females have a larger uterus that will accommodate more pups.

Within these numbers lie some interesting examples. Sand Tigers produce two young per reproductive cycle. That number isn't arbitrary. A female shark has two uteruses. Several fertilized eggs are released into each uterus, but the first one to hatch internally is destined to be the only one to survive. The firstborn Sand Tiger devours all its uterine siblings, engaging in what is called intrauterine cannibalism. After finishing off its brothers and sisters, it then feeds on the eggs that the mother continues to produce. At the end, only two well-fed, meter-long babies emerge, one from each uterus. Sharks are the only animals known to engage in intrauterine cannibalism.

Most live-bearing elasmobranchs have low average birth rates, producing fewer than a dozen young each breeding cycle, which is seldom less than a year and often longer. Intermediate numbers characterize a smaller set of species, such as Nurse Sharks and Tope. A few species are more prolific breeders; these include the Prickly Sharks, hexanchid cow sharks, Tiger and Blue sharks, and Whale and Pacific Sleeper sharks.

Sharks: The Animal Answer Guide

Scalloped Hammerheads produce intermediate numbers of young compared to most other shark and ray species. Shown here is a female Scalloped Hammerhead and her 13 fully developed pups that were removed from her uterus after capture. Each pup still has its umbilicus attached. Photo by Charles Stillwell, National Marine Fisheries Service, National Oceanic and Atmospheric Administration

Egg-laying selachians produce intermediate numbers. The 10 to 50 eggs of heterodontid Horn Sharks are spread over a one- to two-month laying season. Scyliorhinid catsharks lay pairs of eggs at varying intervals and numbers.

Rajiforms are egg-layers and fall into the intermediate production category. Little Skates spawn twice a year and produce up to 35 eggs annually. Blue Skates, the world's largest skate at 285 cm (9.5 ft) long and 100 kg (220 lb), lay only 40 eggs annually. An exception to these numbers is the Mottled Skate (*Raja pulchra*), one of the few species that encases more than one young in an egg case. Females produce an average of 240 fertilized eggs per year, but the number can reach 550. In contrast, live-bearing batoids such as guitarfishes and stingrays produce only about six young per year, following the low-output pattern seen in live-bearing versus egg-laying selachians.

Another interesting aspect of shark reproduction is the state of development of young in a female. In many species, a mother shark may be carrying young of varying developmental stages, from undeveloped eggs right through to pups ready to be born. Some embryos inside Nurse Sharks may be 50 to 100 mm (2–4 in) long and still in egg cases. Other, larger embryos (215–290 mm, 9–12 in long) are free from their egg cases, the smaller still attached to yolk sacs and the larger ones having absorbed all their yolk, having only yolk sac scars. Whale Shark mothers may also be carrying young in various stages of development, including 42-cm-long (17-in) embryos still in egg cases and 64-cm-long (26-in) mobile young close to being born.

To the best of our knowledge, live-bearers release all their young at about the same time. However, many egg-laying species deposit a few eggs at a time, spreading egg laying out over several days. Horn Sharks and skates deposit a few eggs every few days over several months, with a total egg output

of 10 to 50 or more. However, some scyliorhinid cat sharks hold onto their eggs until many are ready to be laid and then deposit them all at once.

Chimaeras, all of which are egg layers, also spread egg laying out over an extended period. Chimaeras are thought to lay two egg cases at a time, one from each oviduct, repeating the process several times during the laying season.

How often do sharks reproduce?

The frequency of reproduction reflects two different aspects of reproductive biology—the gestation period plus the "interbirth interval," or the time between bouts of reproduction. Together, they tell us how much time a female needs to replenish the fat stored in her liver before mating and producing new young. Many live-bearing sharks reproduce on an annual cycle and require relatively little time for the females to recover before becoming pregnant again. Atlantic Sharpnose, Smooth Dogfish, Scalloped Hammerhead, and Pacific Angelsharks are examples, producing young each year. They do not appear to have a resting period, mating again shortly after they give birth. Most rays also reproduce on an annual cycle.

But in many other elasmobranchs, large gaps exist between one reproductive cycle and the next. For example, manta rays give birth to a single pup after a year-long gestation period, but they seldom bear young in consecutive years. Field observations suggest that they rest for at least a year and maybe for as long as six years before reproducing again. White-Spotted Marble Electric Rays have a one-year gestation period followed by two nonpregnant years, as do Sandbar and Dusky sharks. Sawfishes are thought to mate once every two years. If they have a "normal" gestation period of one year or less, they too rest for a year before producing new young. Even the relatively prolific Nurse Sharks, with their five- to six-month gestation period, produce litters only every other year, which means they take a year off between litters, as do Blue Sharks and Broadnose Sevengill Sharks. Blacktip and Finetooth sharks have a nine-month resting period. We know frustratingly little about the breeding habits of White Sharks; it is estimated that they have a gestation period of between 12 and 22 months, with an unknown interbirth interval.

Interbirth interval is another important piece of biological information for shark conservation because it tells us how quickly individuals are replaced in an exploited population. Unfortunately, it is also information that is difficult to obtain.

What is a baby shark called?

The young of many bony fishes are called fry, and some have special names—alevins, parr, and smolt in salmon; elvers in eels; and ammocoetes in lampreys. In addition, many bony fishes have very small larvae that are very different from the adults, and these larvae have their own names. But shark babies are miniature adults, not distinct larvae. Baby sharks are usually called pups.

Are all the baby sharks in a mother full siblings?

Given the variability of reproductive behavior and development among sharks, it isn't surprising that the answer to this question is yes and no. Paternity analysis (identifying the father) has been attempted only for a small number of species, but the results show the complete range of possibilities. Bonnethead Shark pups are usually full sibs, fathered by the same male, although occasionally multiple fathers are involved in the case of larger females that are carrying larger litters. Litters of Lemon Sharks have more than three fathers. Other species analyzed include Spiny Dogfish (two fathers), Nurse Sharks (more than four fathers), Shortspine Spurdog (usually only one), Bignose Shark (more than two), and Sandbar Shark (two to five). A Bluntnose Sixgill Shark that washed ashore near Seattle, Washington, contained 80 embryos. DNA tests showed that eight different males had fathered the babies, thus far a record number among elasmobranchs. Tagging studies suggest that although a litter may have several fathers, littermates remain together in Puget Sound for extended periods.

Even egg-layers appear to mate with several males each cycle. More than 90% of female Small-spotted Catsharks studied had mated with several males.

This small sample of species suggests that many if not most shark females mate with several males during each breeding season. One strong exception to this generalization is the Whale Shark. Twenty-nine embryos taken from a litter of 304 found inside one female Whale Shark showed a complete range of developmental stages and sizes. Despite this range, all the young were full sibs sired by the same father. This would indicate that female Whale Sharks can store sperm and produce eggs over a long period, allowing the eggs to be fertilized at different times.

Do sharks care for their young?

Yes, but only before birth. No sharks, skates, rays, or chimaeras are known to care for their young once a baby is born or an egg is laid. The actual "care"

provided by most chondrichthyans involves carrying a developing embryo inside its body, safe from predators. The mother shark also provides oxygen, water, and minerals for a developing embryo, while carrying away metabolic waste products in her bloodstream. But pre-birth or pre-laying care mostly involves providing nutrition to the embryo (see "Do sharks lay eggs or do they give birth to live young?" above).

If we include egg shape and egg-laying behavior, we can expand the list of care-giving species. Some chondrichthyan eggs have shapes and structures that increase their invulnerability, or are deposited in places that predators can't reach. Horn Sharks have especially complex egg shapes, the leathery shell forming a spiral that the mother pushes into cracks, basically corkscrewing it down and out of harm's way and making it very hard to extract from a crevice. Some skates in the genus *Sympterygia* (*S. acutus* and *S. bonapartei*) along the Brazilian coast produce egg capsules with very long tendrils. Females lay their eggs in bunches, the tendrils attached to debris and intertwined by the female's egg-laying movements, forming a "nest" that looks like a bouquet of flowers and could help protect the developing young.

How is the sex of a shark determined?

The sex, or gender, of an animal can be fixed at fertilization by inherited genes or can be affected instead by environmental factors such as temperature. In most "higher" vertebrates (birds and mammals), sex is controlled by genes from the start, whereas in many lower vertebrates (some bony fishes, turtles, alligators), environmental and even social factors such as crowding can cause an animal to be one sex or another and even to change sex during its lifetime. It appears that sharks follow the higher vertebrate plan, their sex being decided genetically when sperm and egg fuse. Males and females in even the smallest newborn elasmobranchs can be told apart by definite (albeit small) claspers on males and by their absence in females. Internally, immature males and females both already have identifiable structures that will turn into testes and ovaries. Sex chromosomes, usually similar to the XY male and XX female pattern of humans, have been found in skates, guitarfishes, stingrays, and many but not all sharks. Sex-determining factors probably reside on these chromosomes.

Exceptions do arise occasionally, suggesting that the "instructions" coming from sex-determining genes can be misunderstood. A Little Skate captured off the New York coast possessed a mixture of male and female reproductive parts. On its left side it had a large, well-developed testis, vas deferens, and functional clasper. On its right side, the clasper was tiny, and internally the skate possessed an ovary, a shell gland, and an oviduct.

Horn Sharks, *Heterodontus francisci*, **produce complex leathery egg cases that have a conical spiral shape. The mother shark literally screws the egg case into cracks in the sea floor, making it more difficult for predators to remove.** Photo by devra, Wikimedia Commons, http://en.wikipedia.org/wiki/File:Horn_shark_egg_morro_bay.jpg

Can sharks reproduce asexually?

Normal reproduction in chondrichthyan fishes involves males and females, with the male's sperm fertilizing the female's eggs. Asexual reproduction, also called "parthenogenesis," or virgin birth, involves a female producing eggs that develop into embryos without being fertilized by a male.

We have yet to find a chondrichthyan in the wild that appears to have been produced via asexual reproduction. However, captive female sharks have given birth to young without mating with a male; genetic testing ruled out the possibility that the female had mated in the wild and stored sperm. At least four examples of parthenogenesis are known in sharks (but not in rays or chimaeras). In 2001, a Bonnethead Shark was born without a father at the Henry Doorly Zoo in Omaha, Nebraska. The mother had not been in contact with a male for over three years, and genetic testing confirmed that no male genes were present in the newborn (which was unfortunately killed by a stingray in the same tank shortly after it was born). In 2002, a female White-spotted Bamboo Shark at the Belle Isle Aquarium in Detroit, Michigan, laid fertile eggs from which two babies hatched. The mother had been in captivity for six years and had never been in the presence of a male. In 2008, a female Blacktip Shark died during a routine physical exam and when opened was found to contain a single, well-developed pup that was ready to be born. Genetic testing again confirmed no genetic material from a father. And between 2007 and 2011, a captive Zebra Shark at the Burj Al Arab Hotel ("the World's Most Luxurious Hotel") in Dubai, gave birth to 15 pups from eggs laid in the aquarium. DNA analysis again confirmed that the female had not mated with a male.

It is tantalizing to speculate about these multiple events. Does parthenogenesis occur in the wild? Or are these virgin births somehow a result of

the artificial conditions—including a lack of male companionship—that the females experienced in captivity? Many skates and rays are kept in captivity, but none have, as yet, reproduced asexually. Virgin births are another aspect of our limited knowledge of the reproductive ecology of elasmobranchs.

Is there such a thing as a hybrid shark?

Hybrids are offspring that result when two different species mate. Until 2012, no one thought sharks or their relatives hybridized. (We can ignore all the monster movies in which mad scientists mate sharks with a variety of other animals to terrorize the world's teenagers.) But a study of mitochondrial DNA in two blacktip species from the east coast of Australia—the Australian Blacktip (*Carcharhinus tilstoni*) and the Common Blacktip (*C. limbatus*)—has changed our thinking. Researchers showed that a large percentage of individuals identified as Australian Blacktips because of their outward appearance were in fact hybrids between the two species. Aside from being a spectacular "first," this result has implications for fisheries conservation, our understanding of evolution in sharks, and the possible response of species to global climate change. Importantly, if hybrids are misidentified as Australian Blacktips and are less fit than full species, fishery quotas for the real species could be set too high, potentially leading to overfishing.

Do male and female sharks live together?

Except during the breeding season, many sharks tend to segregate by sex, males occurring predominantly with males and females with females. Among the species known to form such relatively sex-specific aggregations are Spiny Dogfish, Scalloped Hammerheads, Small-spotted Catsharks, Oceanic Whitetips, School Sharks, Sand Tigers, Gray Reef Sharks, Bonnetheads, Blue Sharks, Cownose Rays, and Thornback Rays (*Raja clavata*). For example, Shortfin Mako males and females occupy different parts of the Pacific off of Japan for most of the year. Females occur widely in subtropical and temperate regions, whereas males are more narrowly distributed in southeastern portions, below 30° north latitude and east of the International Dateline at 180° longitude. Females also show a seasonal distribution: when they are pregnant, their distribution shifts from southerly latitudes to more northerly latitudes as their embryos grow.

Sexual segregation applies to other migratory species. We now believe there is one wide-ranging world population of Whale Sharks, suggesting that animals come from many places to breed together, perhaps in a few, still-unknown sites. But at other times they separate by size and sex. Aggregations of large females have been observed in only two locales: at the

Sharks: The Animal Answer Guide

southern end of the Gulf of California near La Paz and in the Galapagos Islands. The known feeding aggregations at Ningaloo Reef off western Australia, in the Maldives, at Gladden Spit off Belize, and off Djibouti near the Arabian Peninsula contain mostly males, most of which are juveniles.

Some chimaeras also form sex-specific, or at least sex-dominant, groups. Of 62 individual Rabbitfish caught in a small area of the North Atlantic, 82% were females, mostly of adult size. The Spotted Ratfish also segregates by sex; males and females occur in separate groups, with juveniles of both sexes aggregating in deeper water than adults. Australian Elephant Fish show the opposite pattern, younger fish occurring in shallower areas than adults. Sex segregation in chimaeras may be a general trend, but we know details only about the species that occur in shallower areas accessible to sampling and fishing gear.

The occupation of shallower water by younger fish, regardless of sex, is a distribution pattern more typical of fishes in general. This pattern probably applies to many shark species, especially the ones that use nearshore pupping and nursery areas (see "Where do sharks lay their eggs or give birth?" above).

At what size and age do sharks mature?

Male and female elasmobranchs can be told apart almost at birth because of a male's claspers (and a female's lack thereof). However, the claspers of an immature male are not very impressive, which is to say they are pretty small. When the male matures, there is sudden and obvious clasper growth, something we can tell just by looking at the outside of the animal. The claspers also become calcified (hardened). For females, it's more difficult to pinpoint the size at maturity. Accurate determination usually requires dissecting the reproductive tract and finding developing eggs. But we have a good idea for many species. In some species, males and females mature at about the same ages and sizes; in many others, one sex matures smaller and/or younger than the other.

Sizes at maturation vary greatly among sexes and locales. In general, females mature at larger sizes and greater ages than males. For example, Scalloped Hammerhead males mature at 140 to 165 cm (5.5 ft), females at 210 cm (7 ft). Basking Shark males mature at 450 to 600 cm (15–20 ft), with females maturing between 700 and 1,000 cm (23–33 ft). Shortfin Mako Shark males mature at 200 to 215 cm (6.5–7 ft), while mature females are much larger, at 275 to 290 cm (9–10 ft). Batoids and holocephalans show similar patterns.

Ages at maturity vary in species depending on sex and also as a function of where the sharks occur. In species in which the sexes mature at differ-

Size and age at maturity and longevity of some sharks, skates, rays, and chimaeras

Species	Common name	Size at maturity (cm)[a]	Age at maturity (yrs)	Maximum known age (yrs)
Selachian sharks				
Alopias vulpinus	Common Thresher	226–400	5–6	22
Carcharhinus leucas	Bull	180–230	21	32
Carcharodon carcharias	White	360–500	9–16	22 (30?)
Carcharias taurus	Sand Tiger	220–230	6–10	17
Centrophorus squamosus	Leafscale Gulper	110–158	45	70
Galeocerdo cuvier	Tiger	250–350	7–11	27
Galeorhinus galeus	Tope	120–185	13–25	43
Ginglymostoma cirratum	Nurse	230–240	16–25	35
Isurus oxyrinchus	Shortfin Mako	200–290	25	32
Lamna ditropis	Salmon	180–220	3–10	25
Prionace glauca	Blue	170–220	4–6	16
Rhincodon typus	Whale	800–900	22–30	27
Sphyrna lewini	Scalloped Hammerhead	140–210	4–9	19
Sphyrna mokarran	Great Hammerhead	234–300	—	44
Squalus acanthias	Spiny Dogfish	52–120	15–35	100
Triakis semifasciata	Leopard	70–130	7–10	24
Batoid skates and rays				
Dasyatis americana	Southern Stingray	—	5–6	28
Dipturus batis	Blue Skate	160+	11	23
Manta birostris	Giant Manta	380–460	—	—
M. alfredi	Reef Manta	270–300	10–15	22–32+
P. microdon	Largetooth Sawfish	240–400	10	30–44
P. pectinata	Smalltooth Sawfish	—	10	30
Raja rhina	Longnose Skate	70–100	10–11	17
Squatina californica	Pacific Angelshark	86–108	11	35
Torpedo californica	Pacific Electric Ray	73+	9	16
Urolophus lobatus	Lobed Stingaree	28+	4	14
Holocephalan chimaeras				
Callorhynchus capensis	St. Joseph/Cape Elephantfish	50+	3–4.2	10
C. millii	Elephant Fish/Ghost Shark	70+	3–4.5	10
Chimaera monstrosa	Rabbitfish	46+	11–13	30

Note: The scientific literature on these topics varies greatly depending on source, so the numbers should be viewed as approximate.

[a] To convert centimeters to feet, multiply by 0.033.

ent ages, males tend to mature at younger ages. White Shark males mature between 9 and 10 years old, females between 14 and 16 years. Male Nurse Sharks in Florida mature at around 16 years old, whereas females don't mature until they are about 25 years old. Male Sand Tigers in the North Atlantic mature at 6 to 7 years old, but females don't reach maturity until they are 9 to 10 years old.

Striking differences also occur in species that have wide geographic ranges. Salmon Sharks in the northwestern Pacific mature at 5 years (males) and 8 to 10 years (females), whereas males of the same species in the northeastern Pacific mature as young as 3 years old and females at 6 years old. Spiny Dogfish (*Squalus acanthias* and *S. suckleyi*) occur around the world at temperate latitudes. The average age at maturity for *S. acanthias* females is reported as 7 to 8 years off eastern Canada, 9 to 11 years off the British Isles, and 12 years along the U.S. Atlantic seaboard. *S. suckleyi* matures at even greater ages: 20 years in the northwestern Pacific and 35 years along the British Columbia coast.

The take-home message from these comparisons is that sharks and their relatives are slow to reach maturity. On average, female sharks don't mature until they are 17 years old, and batoids don't mature until they are about 8 years old.

How long do sharks live?

Compared with bony fishes or small mammals or birds, sharks and their relatives tend to live a long time. The average longevity for larger sharks is 35 years, and for batoids 26 years. Chimaeras don't appear to live as long as sharks or rays, but again we have age information on only a few species, and these species may not be representative of the group as a whole.

Looking at particular species that many people are interested in is a little frustrating. For example, we are not sure how long White Sharks live, but we do know they can live as long as 22 years and maybe as long as 30 years. That first number comes from growth rings on the vertebrae of dead sharks. Supporting information comes from photographs of the dorsal fins of five White Sharks off central California. These animals were seen over a 16- to 22-year time span. Sharks have the same marks and scars for that long, which allows researchers to identify individuals. Dorsal fin marks were also used to establish that Nicole—the female tagged off South Africa, tracked to western Australia, and then rediscovered back off South Africa—was the same animal (see "Do sharks migrate?" in chapter 5). Such data also demonstrate that the same sharks return to the same locales year after year. Longevity information for Reef Mantas is similarly limited to sightings of known individuals. One female was seen repeatedly in the Maldives for 22 years and was mature when first observed, making her at least 32 years old. Information on observed maximum ages for a variety of Chondrichthyes is presented on page 130.

We don't know how long Basking Sharks live because we don't have a reliable way of measuring their ages (see the end of the next question).

How fast do sharks grow?

Sharks grow throughout their lives, rapidly at first, then at slower and slower rates as they age. Taking the White Shark as one example, a new-born White Shark female is about 1.5 m (5 ft) long. At two years old, she is about 2 m (6.5 ft) long, having increased 33% in one year. By the time she reaches maturity at around 5 m (16.5 ft) long, she is likely to be a 14-year-old teenager. During the initial growth phase, White Sharks feed primarily on fish and grow slowly. Growth accelerates when the teeth become triangular and the diet often changes to calorie-rich marine mammals, which occurs at between 3 and 4 m (10–13 ft) in length. In large adult White Sharks, growth slows considerably, an animal increasing by only about 6% of its total body length per year, although weight and girth may increase at a greater rate.

Growth in different species will vary by locale, sex, stage of development, and diet. Growth rates are known for a number of chondrichthyans, the results usually reported as percentage increase in body length each year. Evaluating these and other available numbers, we can conclude that male chondrichthyans grow at an average rate of about 20% per year, and females grow at the same or a slightly lower rate, with many exceptions.

How can you tell the age of a shark?

Researchers study the few hard parts of sharks to determine their ages. Growth rings of calcified cartilage and proteins are laid down regularly on hard structures such as fin spines on Spiny Dogfish, Horn Sharks, and chimaeras, and on tail thorns of skates. Scales can be used to determine the ages of bony fishes but not for sharks because sharks shed their scales regularly. The "bony" barbs of stingrays can't be used for the same reason.

When magnified or x-rayed, thin sections sliced from spines or the backbone of a shark show dark and light bands that represent slow and fast growth in cold and warm seasons, much like tree rings. The dark lines occur where the added layers are close together owing to slow growth, such as in winter. The light lines are the layers that were added much faster, as during the summer. Each pair of rings should represent a year's growth, except in tropical species, which don't necessarily face predictable seasonal periods of slow and fast growth.

The age and growth of a great number of shark, ray, and skate species have been described from the vertebral banding patterns of animals caught at sea. Results are sometimes uncertain because it is difficult to determine if a band pair truly represents one year's growth, something that has to be verified from studies of captive animals or animals that have been caught,

Growth rates of some better-known elasmobranchs

Common name	Species	Males (% per year)	Females (% per year)
Sharks			
Tiger	*Galeocerdo cuvier*	18	18
Blue	*Prionace glauca*	18	13
Scalloped Hammerhead	*Sphyrna lewini*	22	25
Spiny Dogfish	*Squalus acanthias*	20	17
Sand Tiger	*Carcharias taurus*	16	11
Rays			
Pacific Electric Ray	*Torpedo californica*	14	8
Pacific Angelshark	*Squatina californica*	15	16
Smalltooth Sawfish	*Pristis pectinata*	8	8
Longnose Skate	*Raja rhina*	25	16
Lobed Stingaree	*Urolophus lobatus*	51	37
Southern Stingray	*Dasyatis americana*	21	53
Chimaeras			
Elephantfish	*Callorhynchus millii*	23	10

Note: Values are average increase in growth per year, given as a percentage of body length.

An X-ray photograph ("radiograph") of one-half of a vertebra from a 170-cm (5.6-ft), 10-year-old female Salmon Shark from Alaska. The vertebra has been sliced as one would cut a pizza pie in half, with the side view of the left half shown. The pairs of dark and light bands make up the growth rings, each year of growth equal to one band pair. Light bands are periods of fast growth; dark bands, of slower growth. *B*: growth at birth; *PB*: growth before birth; *CR*: *centrum radius*, or length of the section analyzed. *Corpus calcareum* is the anatomical name of the outer region of the vertebra, where banding is most obvious. Photo courtesy of Ken Goldman, from K. J. Goldman and J. A. Musick, "Growth and maturity of Salmon Sharks (*Lamna ditropis*) in the eastern and western North Pacific, and comments on back-calculation methods," *Fishery Bulletin* 104 (2006): 278–92

injected with dyes that stain hard structures, and then recaptured later. Sometimes more than one ring is deposited in a year. One novel verification method involves analyzing museum specimens that were alive during the nuclear bomb testing period of the 1950s and 1960s, when radioactive carbon was taken up by the animals alive at that time. A great deal has been written about these and other aspects of age determination of sharks, most of which appears in technical publications that an interested reader can explore.

The number of bands on vertebrae usually provide accurate counts. But in some sharks, it matters which vertebrae are studied. In most lamnoid sharks—such as Porbeagles, makos, and threshers—all vertebrae have the same number of band pairs regardless of where on the body they are sampled. But older Basking Sharks give different counts depending on which vertebrae are used: vertebrae in the stomach region have as many as 24 more band pairs then vertebrae near the tail (Basking Sharks have about 85 vertebrae). Angel Sharks, Gummy Sharks, and School Sharks also have more growth bands nearer the head than the tail. Researchers have interpreted the differences to mean that bands are added to vertebrae to support the body at that place along the shark, more bands providing greater support. This means that in many sharks the number of bands is not a reliable measure of shark age.

The location on a fin spine where the bands are counted also makes a difference. Chimaera spines give accurate age measures if ring counts are made from the tip of the spine, but counts are unreliable if sections of the base of a spine are studied. Also, fin spines can erode over time, obscuring and even obliterating some growth rings.

Chapter 7

Foods and Feeding

What do sharks eat?

The most general statement that can be made about feeding patterns in sharks is that they are opportunists, eating just about anything. Except plants: sharks don't like brussels sprouts, perhaps attesting to their intelligence. Actually, chondrichthyan fishes don't like any vegetables or fruits. They are as a group entirely carnivorous, eating all types of marine animal life that they can catch, or scavenging on dead animals that other sharks have dismembered. Bony fishes make up 70% to 80% of the diet of most shark species, although a variety of species also eat lobsters and crabs, squid and octopus, porpoises and whales, seabirds, sea turtles, and other sharks. Opportunism means eating what's available. Galapagos Sharks in the Galapagos Islands feed on marine iguanas, a prey type not available anywhere else in the world. But marine iguanas are large, tasty, and slow-swimming.

An example of a shark with a diverse diet, highlighting a species about which we know comparably little, is the Broadnose Sevengill Shark. Among food items found in the stomachs of Sevengills in South Africa have been at least 11 species of sharks; 2 each of skate, stingray, and torpedo ray species; shark egg cases; a chimaera; a hagfish; more than 14 kinds of bony fishes; and squid, octopus, snails, crustaceans, seals, and dolphins. In other places, Sevengills also eat sea lions, rats, and people. The human remains were scavenged from a suicide victim who had jumped off the Golden Gate Bridge.

Scavenging is an important feeding mode in many sharks. Their scavenging habit, coupled with an exquisite sense of smell that allows them to detect baited hooks from great distances, is part of what makes sharks so

vulnerable to fishing. What they eat depends mostly on place and shark species. Among sharks attracted to dead whales and dolphins (and humans) are Sevengill Sharks, Tiger Sharks, and of course White Sharks; dead seals attract White Sharks and Sleeper Sharks; Blue Sharks feed on dead squid, as do some very deep-living species such as Velvet Belly Lanternsharks.

The classic opportunist, in terms of predation and scavenging both, is the Tiger Shark. Tiger Sharks are efficient predators and capture other sharks, bony fishes, crustaceans, mollusks, dugongs, seabirds, sea snakes, dolphins, and sea turtles. But the list of items found in Tiger Shark stomachs reads like an ad for a yard sale: license plates, oil cans, tires, deer antlers, boat parts, jewelry, clothing, tires, books, and baseballs.

Tiger Sharks also take advantage of opportunities afforded by short-term periods of prey abundance. Best studied are Tiger Sharks that feed on albatross chicks in the northwestern islands of the Hawaiian chain. Every June and July, Tiger Sharks arrive at Midway Island and French Frigate Shoals just as Laysan and Black-footed Albatross chicks are making their first attempts at flying. Many of these first flights end quickly in the water, where the sharks are waiting. Tiger Sharks aggregate similarly at Raine Island in Australia to feed on a seasonal abundance of egg-laying sea turtles that become stranded at low tide and die of heat exhaustion (see "How do sharks navigate?" in chapter 5).

Recent discoveries in the Gulf of Mexico have shown that Tiger Sharks feed opportunistically on birds much smaller than albatrosses. Songbirds migrate every spring from South to North America. Those flying across the Gulf of Mexico pass over a string of oil and natural gas platforms. The platforms are lit up at night, and many birds are confused by the lights and circle the platforms for hours until they become exhausted and fall into the sea. This "nocturnal circulation" may involve as many as 100,000 birds circling a single platform. Tiger Sharks have learned about this bonanza. Nearly half of 50 Tiger Sharks caught over a two-year period near these oil rigs contained a birdwatcher's list of small species, including woodpeckers, Scarlet Tanagers, Brown Thrashers, Meadowlarks, catbirds, kingbirds, and swallows.

Tiger Sharks also stalk prey, very much the way that terrestrial tigers do, relying on the element of surprise. Video footage taken via "Crittercams" attached to Tiger Sharks in western Australia show the sharks approaching green sea turtles that are feeding on sea grasses. If the turtle looks up as the shark approaches, the shark veers off. But if the turtle is caught unawares, the shark attacks. Studies of Bengal tigers in India that attacked fishermen in canoes revealed that these predators also attacked from behind. When fishermen wore Halloween masks on the back of their heads, attacks declined significantly. Fishermen who were given masks but didn't wear

Tiger Sharks are well known for their extremely diverse diets, which include live prey, dead animals, and discarded human trash of all kinds.
Photo by Austin Gallagher, http://austin gallagher.com; used with permission

them were attacked at the same rate as fishermen without masks. It's sort of like seatbelts and bike helmets: having one doesn't help much if you don't use it properly. (In case you're wondering, no one has tried painting faces on the backs of sea turtles as a test.)

Some sharks specialize on, or at least prefer, certain food types. Carcharhinid Tiger, Bull, and Galapagos sharks are well known for eating other sharks but are by no means the only ones that do so. Sevengill Sharks have been seen attacking a number of other, smaller sharks, including Leopard Sharks in California and School Sharks in Africa. Sevengills are also cannibalistic. Again, surprising their prey appears to be a tactic of choice for Sevengills. A seemingly sluggish animal, a Sevengill moving slowly and showing no obvious interest in several Leopard Sharks will put on a sudden burst of speed and quickly grab a Leopard Shark, bypassing other, closer Leopard Sharks.

Large hammerheads, such as Great Hammerheads, feed heavily on stingrays. Bull sharks also favor stingrays. Bonnetheads, the smallest hammerhead species, prefer swimming crabs. Sleeper and Cookiecutter sharks take chunks out of marine mammals such as seals and dolphins, although Sleeper Sharks also feed on the ocean's largest invertebrates, the giant and colossal squids. Leopard Sharks, Nurse Sharks, and Green Lanternsharks (*Etmopterus virens*), as well as most skates and rays, feed mostly on invertebrates. Horn Sharks feed preferentially on mollusks such as clams and have the dentition needed to crush hard-bodied prey. Stingrays as a group also possess crushing dentition and often feed on animals with shells.

Foods and Feeding

Although seemingly specialized for certain food types, some sharks can adjust their teeth to take advantage of a meal. Hemiscylliid bamboo sharks can turn their spiky, fish-eating teeth into more pavement-like dentition by allowing the functional row of teeth to bend toward their throat, creating a flat rather than a pointy contact surface (see "Do a shark's teeth change during its life?" in chapter 2).

Some sharks prefer certain types of food over others. For example, White Sharks prefer prey with a high fat content—thus, their attraction to whale carcasses. Tests with floating baits show that White Sharks will bite and reject furry sheep carcasses but will bite and consume the carcass of a nice fat pig. White Sharks offered baits of seal fat and seal muscle will eat the fat baits but reject the muscle. The initial bite (the attack) is apparently when the shark tastes its prey and assesses fat content. This preference for fatty foods may be why furry sea otters are attacked but not eaten, whereas seals and sea lions, with all their blubber, are favored prey items (or maybe White Sharks just don't like a mouth full of fur). A preference for fat has prompted speculation that one of the reasons people are attacked but often not consumed is because we aren't fat enough. An alternative explanation is that we're smart enough to get out of the water, or that someone comes to help before the shark finishes its meal. Needless to say, the controlled experiments necessary to test these ideas have not been performed.

The biggest sharks and rays eat the smallest items. Whale, Basking, and Megamouth sharks and manta rays feed on zooplankton (a collective term for very small animals or larvae that are suspended in the water) and small fishes.

Chimaeras have beaklike teeth. Not surprisingly, the limited information available on their food habits indicates that they crush shelled prey. Diet studies point to hard-bodied, bottom-living invertebrates—particularly clams, scallops, snails, and crabs—as the main prey items. But as good chondrichthyans, they take advantage of a variety of available food types, including small fishes, worms, echninoderms (brittle stars, sea urchins), and even sea anemones and an occasional jellyfish. Smaller holocephalans eat smaller prey or prey types with softer shells.

Finally, given all the strange and seemingly inedible things that have been found in some sharks, you might think that they do not have the ability to reject a food item once it has been swallowed. On the contrary, sharks have a much more effective system for expelling something distasteful than we do. (We rely on stomach muscle contraction and reverse peristalsis in our esophagus, pushing items slowly or sometimes not so slowly back out.) A study of the Shortfin Mako found that these predators can actually evert their entire stomach, basically pushing it up and out of the mouth in as little as three-tenths of a second. They do this by contraction of abdominal

muscles, lifting of the throat region, and additional contractions of the stomach itself. The surrounding seawater then rinses away the rejected items. Afterwards, the stomach is pulled back in through the mouth and back into place by a combination of suction and peristaltic waves passing down the esophagus. Such "gastric oral eversion" is known in several other shark species. Expulsion of objects from the other end of the intestinal tract is accomplished by pushing the intestine out through the cloaca. Between gastric oral eversion and intestinal cloacal inversion, sharks would be able to cleanse and flush their entire gastrointestinal tract. Dogfish may not bark, but they can barf.

How do sharks find food?

Sharks use all five of their senses, plus one or even two we lack, to detect and then home in on food items. Different senses function additively at different but overlapping distances from a food source, increasing the accuracy of detection as the shark closes in on the food.

Hearing and smell operate at the greatest distances and over different time scales. Sound travels fast in water; chemicals diffuse slowly. Numerous experiments have shown that sound alone attracts sharks to a potential food source. Sharks detect sounds with their inner ears, much as we do. Pulsed sounds between frequencies of 20 and 1100 Hz are especially attractive. In the sea, these are the frequencies emitted by a rapidly swimming fish or by the squeaks, squeals, grunts, thumps, and buzzes produced by injured

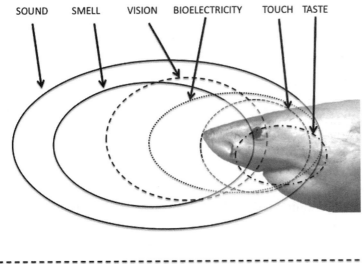

SOUND SMELL VISION BIOELECTRICITY TOUCH TASTE

HUNDREDS+ TENS ONE TO TEN ZERO TO TWO ZERO
DISTANCE IN METERS (1 M = 3.3 FT)

A general picture of the additive use of the senses in foraging sharks. Distances given are approximate and change with water clarity, light levels, prey type and behavior, current speed, and habitat, not to mention shark species. Under some conditions, smell may function at a greater distance than hearing. Illustration by Gene Helfman

fishes. Surprisingly (to us), sharks can tell where those squeaks and thumps are coming from, up to 200 m (660 ft) away (see "How well can sharks hear?" in chapter 2). We cannot locate the source of a sound underwater.

Sharks are often called "swimming noses" because of their incredible sensitivity to olfactory cues (smells). A shark's nose is connected to the very large olfactory lobes of its brain, a sure sign that olfaction is important to sharks. A Lemon Shark can detect fish extracts at levels as low as one part per 25 million parts seawater; Blacktip and Gray Reef sharks react to one part per 10 billion—which is the equivalent of about one drop of fish goop dispersed in an Olympic-sized swimming pool. This sensitivity is greatest to proteins, amino acids, and amines (see "How good is a shark's sense of smell?" in chapter 2).

As with sounds, smells can be detected far from the source, but smells are rapidly diluted and weakened as distance increases. Directional cues can be provided by current flow, with sharks often swimming upcurrent to an odor. An increasing amount of an odor, its concentration gradient, can also provide a directional cue. Differential arrival times at the shark's two nostrils also provide directional information. Hammerhead sharks might be better able to localize an odor source because the wider spacing between their nostrils would increase the time lag between an odor hitting the left and right nostrils as the shark swam at an angle through the odor trail. And the food source doesn't necessarily have to be directly submerged in water. White Sharks have shown up off southern California beaches where government agencies had buried a dead blue whale, much to the vexation of local surfers.

As a shark approaches something it has heard or smelled, vision comes into play, although acuity is obviously affected by water clarity. Sharks have well-developed eyes and large optic lobes in their brains, indicating some reliance on vision. Their eyes are best adapted for viewing objects more than 23 cm (9 in) away and in dim light conditions. Color doesn't matter, at least not to selachian sharks; rays and chimaeras have better color vision (see "How well can sharks see?" and "Can sharks see color?" in chapter 2). Again, the hammer of hammerhead sharks improves their vision by increasing the visual field around them greatly over that of pointy-headed sharks; the wider the hammer, the greater the visual field (see "Why do hammerhead sharks have such strange heads?" in chapter 2).

The final stages of finding food rely on touch, taste, and especially bioelectricity. Sharks often bump nearby objects as a means of testing their edibility. We use touch when we handle food to assess consistency (such as hard/unripe versus soft/ripe versus mushy/rotten fruit). Exactly what information a shark gains isn't clear, and it is possible that the shark is sensing electrical fields as well as consistency (see next paragraph). But sharks

Sharks: The Animal Answer Guide

have another type of touch sense. Sharks can detect minor variations in water movement, often called a "distant touch" sense. Receptor cells in canals along the lateral line and scattered around the body react to minor changes in the flow of water pushed toward a shark, such as those coming off a nearby swimming fish, or water flowing across its body as the shark swims. This distant touch sensitivity includes directional information, which means that as a shark moves through the water, it can detect the presence, location, direction of movement, and relative speed of moving objects that displace water.

Sharks are extremely sensitive to bioelectric fields (see "How do sharks detect electric fields?" in chapter 2). Such fields are created by all living things due to muscular contractions, nerve impulses, or just because the salt concentration of an animal's body fluids differs from the salt concentration of the water around it. Sharks have electrosensory organs, called the Ampullae of Lorenzini, scattered around their bodies but concentrated in the head and mouth region. Another hypothesis about the function of the expanded skull of hammerhead sharks is that the hammer allows for a wider distribution of ampullae, which increases both sensitivity to and accuracy of detection of electrical impulses.

Sharks and rays are so sensitive to electric fields that, theoretically, they could detect the output of a single flashlight battery several miles away if it were not for electrical interference from the earth's geomagnetic field. They are more than sensitive enough to detect the bioelectric output of common prey such as fishes or less common prey such as humans. Sharks can find prey buried in the sand by homing in on the bioelectric output of the prey. Sharks also dive into the sand after buried active electrodes.

When a shark makes its final approach to food, electrical output appears to be the primary cue it uses to direct its bite. When sharks are offered a choice between the fish slurry that initially attracted them (the bait) and a nearby activated electrode, they prefer to bite at the electrode.

Touch and taste are the last senses employed in the search for a meal. Smell and taste are related senses in that both rely on detection of chemicals. Olfactory cues may be used to sense food at great distances, whereas taste buds come into play much closer (very close) to the food item. Taste buds concentrated inside the mouths of sharks apparently play a role in accepting or rejecting food, as when demonstrably hungry sharks reject the Red Sea Moses Sole (*Pardachirus marmoratus*) after biting it. The sole secretes a combination of milky skin toxins called pardaxin and mosesins when attacked. Sharks quickly release a sole but feed actively on other less chemically defended fishes. Sharks also bite into and quickly release hagfish, presumably in response to the hagfish's rapid secretion of mucous, which the shark finds somehow distasteful, if not just repulsive. Noxious

chemicals form the basis of many questionably successful shark repellants that sharks detect via either smell or taste (see "How can sharks be repelled?" in chapter 9).

How do sharks catch their prey?

Because different sharks eat so many different things, the means by which they catch and consume prey vary among and even within species, depending on circumstances. At the most basic level (aside from just biting down on something), sharks can be divided into those that chase after live prey, those that ambush prey, those that inhale prey via suction, and those that filter prey with an open mouth while swimming.

CHASERS. Most sharks probably chase after prey at some time in their lives, if only for a short distance. But those best known for making a living this way are makos, Porbeagle, Salmon, Sand Tiger, Sevengills, and many carcharhinid sharks. Makos are the classic pursuit predator, not surprising because they are also the fastest shark and consequently the most sought-after shark as a game fish (see "What is the fastest shark?" in chapter 1). Many of the sharks that "specialize" in chasing, but especially the makos, Salmon Shark, and Sand Tiger, have long, spiky teeth best adapted for grasping slippery fishy prey on the run.

A special group of chasers make the initial attack using something other than their teeth. Ecologically, these are some of the most interesting feeders, with a diversity of attack methods unrivaled by bony fishes. The devices these sharks use include biological hammers, swords, scythes, and shockers.

We've seen that there are a number of hypotheses about how hammerhead sharks use their "expanded cephalic foil," but one use has been verified by observation. Hammerheads use their flattened head to knock stingrays down to the bottom and then use the same flattened head to pin a ray down while taking bites from its wings.

The Great Hammerhead feeds extensively on stingrays. Snorkeling in 6 m (20 ft) of water over a seagrass bed, researchers in the Bahamas saw a 3-m-long (10-ft) Great Hammerhead chasing a 93-cm-wide (3-ft) Southern Stingray. The shark thrust its head down against the back of the ray, knocking it to the bottom. The shark then pinned the ray against the bottom with its head and pivoted around on the back of the ray, taking a bite from the front edge of the ray's left pectoral "wing." The ray limped off, and the shark pinned it to the bottom again, pivoted, and took a bite from the front of the right pectoral fin. This crippled the ray, which was then eaten by the shark. The shark used its hammer to both knock the ray down and then

The Salmon Shark, a close relative of the White Shark, is found chiefly in the cooler waters of the North Pacific. It feeds primarily on fish, especially salmon—hence, its common name. Both jaws contain the long, spiky teeth characteristic of fish-eating sharks. Photo from www.afsc.noaa .gov/race/media/photo_gallery/photos/ Lamnidae/salmsharkcls.jpg

to immobilize it while taking bites. A pointy-nosed shark probably would have slid off the ray's back, but the hammerhead was able to apply pressure to both pectoral fins at the same time.

The rostral saw of the pristid sawfishes is an impressive weapon. As the defining characteristic of the family, it isn't surprising that the saw has multiple uses depending on circumstances. Although direct observations are surprisingly limited, sawfishes appear to vary the way they use their saw depending on prey type, location, and behavior (ignoring for the moment that the rostrum is studded with electroreceptors that the sawfish probably uses for finding buried prey):

- Sawfishes swing their head back and forth in rapid lateral swipes, stunning or dismembering prey fish. This technique is said to allow them to enter a school of fish and whack a couple. On videos, the back-and-forth swinging is just a blur. Damaged prey (incapacitated and even decapitated) are suctioned up in the water column. If a mullet or herring gets impaled on one of the rostral teeth, the sawfish scrapes it off against the bottom and eats it.
- If the prey dives to the bottom, the sawfish pins it there under the saw, swims forward, and suctions it up.
- Sawfishes reportedly also use the saw as a digging tool to unearth buried crustaceans. Observations of this particular behavior are rare and anecdotal, and the most thorough study of feeding in any sawfish species, using juvenile Largetooth Sawfish, did not observe this activity. Sawfishes do defend themselves readily with the saw, including striking small boats that follow too closely and persistently in shallow water.

Foods and Feeding

Use of the hammer during feeding by a Great Hammerhead Shark. (A) Shark chases ray. (B) Shark strikes down across the back of ray with the flat underside of its hammer. (C) Ray bounces off the bottom from the force of the blow while shark brakes with its pectoral fins. (D) Shark delivers a second downward blow across the back of ray. (E) Shark pivots while holding ray against the bottom and takes a bite from front of left pectoral fin. (F) Injured ray attempts to swim off, followed by shark. From W. R. Strong, F. F. Snelson, and S. H. Gruber, "Hammerhead Shark predation on stingrays: An observation of prey handling by *Sphyrna mokarran*," *Copeia* 1990, 836–40; used with permission of the American Society of Ichthyologists and Herpetologists

If the sawfish does use the technique described above for catching schooling prey, it is one of the few animals that is minimally bothered by the confusion created by a large group of prey fishes, which is a primary reason prey fishes form schools in the first place. The sawfish is freed from the necessity of singling out an individual fish; it can enter a dense school and slash back and forth.

Finally among our group of special chasers, we have thresher sharks, whose tail may account for one-half of the body length. For years, researchers and fishers speculated that threshers slashed at and incapacitated small fish with their scythelike tails, basically using their tail in much the same way that a sawfish is supposed to use its snout. Anecdotal evidence includes many examples of threshers caught by the tail on baited hooks and trolled lures, something that rarely happens with other sharks.

Video evidence has finally appeared that confirms the use of the tail in feeding. Researchers working off the California coast towed video cameras ahead of baits and filmed Common Threshers approaching the baits.

Sharks: The Animal Answer Guide

Nearly half of the 33 sharks filmed struck at the bait with their tail, hitting the bait 65% of the time. Six sharks struck more than once; one individual struck at the bait seven times in less than a minute. An average strike lasted two-tenths of a second, the shark striking the bait about midway down its tail. None of the sharks tried to bite at the bait before striking it with its tail. Other researchers filmed Pelagic Thresher Sharks (*Alopias pelagicus*) near the Philippines feeding on schools of sardines. A shark would rapidly enter a school of prey, brake with its pectoral fins, and sling its tail tip over its head, literally whipping the tip at the sardines. The tip can reach speeds of 129 kph (80 mph) and injured the prey, sometimes breaking their spines and even cutting them into pieces. The disabled prey were then swallowed. These observations clearly show that tail strikes are a common and even preferred feeding tactic, at least in these species.

AMBUSHERS. Ambush predators are frequently referred to as "lie-in-wait" foragers, evoking the image of a lioness hiding in tall grass and pouncing on a passing zebra or tourist. Sharks and their relatives seldom have the convenience of tall grass to hide in. Instead, they either move slowly through the water, sneaking up on prey, which they capture with a sudden burst of speed, or they lie partially or completely buried in bottom sediments, lunging out at passing fish and crustaceans. The element of surprise is key to success in ambush predators.

Some of the largest sharks, including White Sharks and Tiger Sharks, are ambush predators of the first type. White Sharks have been extensively photographed launching themselves up into the air, a squirming young Cape fur seal tightly grasped in the deadly jaws. A typical feeding sequence near Seal Island, South Africa, begins shortly after dawn, with the shark moving slowly near the bottom. Its dark back blends with the dark bottom below, the camouflage enhanced by the dim light conditions of early morning. Upon detecting the silhouette of an unsuspecting seal at the surface, the shark explodes to the surface, snapping its jaws closed as it engulfs the seal. The surface rush has been named the "Polaris breach" because it is reminiscent of a Polaris missile being fired upward from a submarine. The success rate of initial breaches and lunges observed at Seal Island was about 31% and declined rapidly if the seal escaped the initial attack; fewer than 20% of feeding sequences involving multiple capture attempts were successful. (Some individual sharks, however, are successful as much as 80% of the time, and White Shark attacks on sea lions off the Farallon Islands near San Francisco are successful 99% of the time.) Once the shark loses the element of surprise, the likelihood of success plummets. An average attack event is over in less than a minute and a half.

Tiger sharks feed on all sorts of active prey, including dolphins, sea tur-

tles, dugongs, sea snakes, other sharks, and bony fishes. Despite their reputation for ferocity, they seldom chase after prey, preferring instead to sneak up on it. As described earlier in this chapter ("What do sharks eat?"), if the prey detects the approaching shark, the shark veers off. The Tiger Shark's overall foraging strategy is largely one of stalk and surprise. Spear fishermen in Hawaii say that you don't have to worry about a Tiger Shark that you've seen; it's the ones you don't see that are likely to attack. We're not sure that's comforting.

Classic ambush predators, those that lie concealed on the bottom and explode upwards to snag prey, exist among both rays and sharks. For example, a South African butterfly ray species, the Diamond Ray (*Gymnura natalensis*), lies on sandy bottoms, where its light coloration helps camouflage it. When a spawning squid approaches closely, the ray lunges out and engulfs it.

The best-known ambushers among sharks are the squatinid angel sharks, a crossover group that is a shark with the body of a skate. (The giveaway clue is that angel sharks have gill openings on the sides of their body rather than underneath, and the pectoral fins are not attached to the head as in batoids.) Angel sharks typically find an ambush site near a reef, lightly cover themselves with sand, and wait. If a prey fish or a squid swims by close enough (no farther than 15 cm, or 6 in), the angel shark erupts from the bottom at an upward angle, rapidly opening its mouth, which creates tremendous suction pressure. At the same time, it protrudes its jaws, thereby inhaling and impaling its prey, facilitated by extremely sharp, pointed conical teeth and strong jaws. An entire strike sequence takes less than one-tenth of a second.

The electricity-generating torpedo rays are good examples of sharks whose hunting behavior varies with circumstance. During daylight, torpedo rays lie in wait in the sand in the fashion of an angel shark. When prey wander nearby, the ray leaps up and over the prey, cups its pectoral fins (where the electric organs are located) around the prey, discharges its electricity, and then swallows the disabled victim headfirst. Output from the electric organs can reach 45 volts and one kilowatt at a rate of 150 to 200 pulses per second. But at night torpedoes swim or drift slowly through the water in search of prey that rest just above the bottom. When a prey fish is encountered, the ray approaches to within 5 cm (2 in), lunges forward, and wraps its pectoral fins around the fish. It then discharges its electric organ, stunning the prey, sometimes breaking the prey's spinal column by the force of the electric shock. The immobilized prey is then grasped in the mouth or forced to the bottom and consumed.

One other type of ambush predator is known, or at least hypothesized, among sharks. These are the frequently mentioned 50-cm (20-in) dalatiid

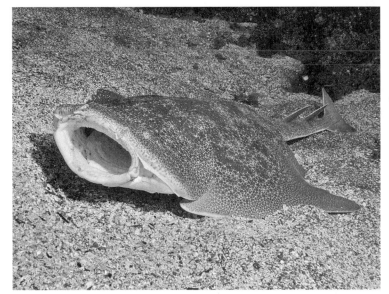

A Japanese Angel Shark (*Squatina japonica*) lies on the bottom, with its mouth partially open. Angel sharks frequently bury most of their bodies in the sand, making them more difficult to detect. They can erupt from the sand quickly, whereupon they literally suction prey out of the water column, helped by a large mouth and long, spiky teeth that prevent the prey's escape. Photo by Ryo Sato, Wikimedia Commons, http://en.wikipedia.org/wiki/File:Squatina_japonica2.jpg

Cookiecutter sharks that feed by gouging round plugs of flesh from larger tunas, whales, dolphins, swordfish, Megamouth Sharks, and swimmers. (See "What kind of teeth do sharks have?" in chapter 2 and "Are sharks dangerous to people or pets?" in chapter 9.) Cookiecutters are among the few bioluminescent sharks, having light-emitting photophores (light cells) along their bellies. It is thought that their photophores help them blend into the background of dim lights produced by other small organisms in an otherwise black ocean. When larger predators swim down to feed on small fishes and invertebrates, the Cookiecutters, who have camouflaged themselves among these small animals, lunge out and remove a plug of flesh. It has even been suggested that the Cookiecutter's lighting could lure its "prey" to within striking distance.

SUCTION FEEDERS. Most sharks, skates, and rays (but not chimaeras) suction food into their mouths. For some, it is the primary means of capturing prey, especially among the heterodontiform horn sharks, orectolobiform carpet sharks, and Nurse Sharks, and a number of batoid skates. Sharks generate suction (negative) pressures by rapid "buccopharyngeal expansion"—pushing down on floor of their mouth and throat, thereby expanding the mouth cavity. Unlike bony fishes, sharks cannot produce much suction by shooting their mouth out, although doing so does bring a shark, or at least its mouth, closer to a food item. Phil Motta and his associates found that in Nurse Sharks, the negative pressures generated during suction can be as high as one atmosphere (14.7 pounds per square inch), "the greatest subambient pressures reported for an aquatic-feeding vertebrate"

(*Journal of Morphology*, Sept. 2008, p. 1041). As powerful as this may be, the pressures fall off rapidly with distance, and it is unlikely that prey can be inhaled from a distance much greater than about 3 cm (1.2 in).

Many sharks use suction to capture buried or hidden prey. Leopard Sharks suck worms out of their burrows. Hemiscylliid sharks such the Epaulette Shark and the Whitespotted Bamboo Shark shove their heads into bottom sediments far enough to cover the first gill slit and suction up worms and crabs. They then winnow the prey from the sediment by swishing the mixture around in their mouths. The prey are then swallowed while sand and silt are expelled out the gills.

Batoid rays use suction as much as or more than selachian sharks do. The Lesser Electric Ray (*Narcine bancroftii*) specializes in wormlike prey, apparently including small, eel-like fishes. Instead of sticking its head into the sand, it protrudes its jaws—which when extended are as long as its head—deep into bottom sediments and suctions up prey (and sand). The mixture is winnowed in the mouth and the sediment expelled out the mouth, spiracle, or gill slits.

Suction can be applied to the task of feeding in other, unexpected ways. Cownose Rays employ "hydraulic jetting." They sit on the bottom, balanced on the tips of their pectoral fins. They then repeatedly open and close their mouths three times per second, drawing water in and shooting it back out. The jet of water thus created is directed at the bottom, where clams and sand are suspended. The sand is progressively flushed away by the stream while the heavier clams remain (see also "Do sharks use tools?" in chapter 4).

The ability of many batoids to use suction pressure in their feeding has even led to the naming of a new species. A rare electric ray from South Africa, the Ornate Sleeper Ray (*Electrolux addisoni*), was known to scuba divers before one was captured and officially described in 2007. Divers had videotaped the animal and showed that it feeds by moving slowly along the bottom, protruding its jaws and inhaling worms by means of a "vigorous sucking action," reminiscent of a vacuum cleaner.

FILTER FEEDERS. Large filter feeders like Whale, Basking, and Megamouth sharks and manta rays don't need much stealth or speed to capture the zooplankton—small floating animals—that they eat. The challenge isn't so much outrunning the prey as it is keeping it from slipping past the filter. Each one of these giant filter feeders accomplishes this in a slightly different manner, with food captured on gill rakers, which are inward-facing, fingerlike extensions from the gill bars that support the gill filaments involved in breathing. Technically, these structures are referred to as "filtering pads," and they differ in each species. Basking Sharks have bristlelike

pads; Whale Sharks have widely spaced, flattened filtering pads; mantas possess rigid, leaflike, folded pads; and Megamouth Sharks have furlike, short gill rakers.

Basking and Whale sharks and manta rays swim with open mouths, employing what is commonly called "ram feeding." This means that water is forced through the mouth and out the gill slits by the forward motion of the animal, an action that serves to continuously trap zooplankton on the filtering pads. All swim slowly, between 0.5 and 1 m/s (roughly 1–2 mph). Megamouth Sharks rely more on suction to draw water in, which is then forced back by closing the mouth. Filter feeding has also led to a very basic change in the mouth location compared with most sharks: Megamouth and Whale sharks and manta rays are among the half dozen chondrichthyans whose mouth is located at the front of the head rather than below it.

Basking Sharks feed as individuals at low plankton densities, but when they encounter high concentrations of food, they form lines of feeding sharks. Each shark drops and spreads its lower jaw to create a circular hoop that can be a meter (3 ft) across. The shark swims slowly, thus posing for what has become the classic Basking Shark feeding photo (usually taken of sharks feeding at or just below the surface). Plankton is caught on the bristlelike pads. The bristles have a microscopic, 0.8-mm (0.03-in) gap between them that traps food but allows water to pass. The shark closes its mouth, backflushes the gill rakers, and then swallows.

Whale Sharks also feed individually or in groups near or at the surface. They have a more varied feeding repertoire than Basking Sharks. They may ram-feed much like baskers while swimming slowly, at about the same speed as baskers, but with the upper jaw out of the water. They can also stop and suck water in and out in pulses, the method they commonly use when fed in captivity (such as in very large public aquariums). An additional technique involves rising vertically out of the water and then sinking quickly back, making water and prey flow into their open mouths. Food is caught by mesh-covered pads that fill the mouth and throat, attached to the inside of the gill filaments. Whale Sharks occasionally "cough" as an apparent way to backflush and then swallow organisms and eggs trapped on the pads.

Mantas are the most active of the plankton feeders, foraging along the bottom as well as up in the water column. They appear to fly through the water, slowing as they encounter patches of prey, twisting, turning, and somersaulting ("barrel rolling") backwards to stay in concentrated patches. Unlike the others, mantas will also feed just above the bottom. They funnel plankton-bearing water into their mouths and out their gill slits by extending their "horns" in front of them. The horns are forward extensions of their pectoral fins, making them the only vertebrates with three pairs of paired limbs. Food is trapped by mucous that covers modified toothlike

Whale Sharks feed primarily by swimming through the water with their mouths open. Water passes into the mouth and out the gill slits. Food items in the water flow into the mouth and are trapped on the filtering pads associated with the gill rakers. At the bottom left, one entire gill arch has been dissected from the right side of the shark. The filtering pads are the darker structures on the right of this inset; the gill filaments used in breathing are lighter and to the left. Redrawn and used with permission of Phil Motta from an original copyrighted by Emily S. Damstra

denticles on each gill raker. Trapped food is then moved to the throat by tiny beating hairlike structures called cilia or by flushing of water across the denticles. When groups of more than 30 mantas gather in areas of high plankton density, they may line up and circle in a head-to-tail chain, referred to as cyclone feeding (see "Do sharks form schools?" in chapter 4).

Megamouth Sharks are rare deep-water animals that have only occasionally been observed in healthy condition, so their feeding behavior is inferred from their anatomy. They apparently engulf zooplankton much like humpback whales, sucking in prey and water, a movement which expands their mouth and throat. The engulfed prey and water is pushed across the gill rakers as the mouth is closed and compressed.

When do sharks feed most?

Most sharks are true opportunists, feeding whenever a meal can be found. But in general, sharks are more active at night than day. Increased nocturnal swimming has been reported for water column species such as Gray Reef, Blue, and Scalloped Hammerhead sharks, but also for bottom-living or resting species such as Port Jackson, Reef Whitetip, Horn, angel, and swell sharks, as well as torpedo rays (see "Are sharks social?" in chapter 4).

As a result, shark fishing is most successful at night, although many "nocturnal" sharks will also take baited hooks during the daytime. Nocturnal sharks often start being active at dusk and return to a resting area at dawn, making evening twilight a period of greatly increased activity. For example, telemetered Lemon Sharks doubled their swimming speed during evening twilight as compared to their daytime or nighttime swimming

Sharks: The Animal Answer Guide

speeds. However, White Sharks are primarily daytime feeders, at least where their major prey consists of day-active marine mammals.

In some species, the difference in activity and location is quite predictable. Gray Reef Sharks at Rangiroa Atoll in the Pacific were inactive in a small lagoon area by day but swam outside the lagoon via a pass in the reef at about the same times each evening, returning to the lagoon via the same route in the morning. One tagged individual exited the lagoon between 6:40 and 6:50 on three consecutive evenings. Sunset and decreasing light are relatively constant at tropical latitudes, and the shark was likely responding to these predictable cues.

How often must sharks feed?

How often and how much a shark must eat depends on species, age, and activity levels. Overall, juvenile sharks need to feed more often than adults. Relatively inactive sharks, such as adult Nurse Sharks, consume about 0.2% to 0.3% of their body weight each day and take about six days to digest an average meal. More active species such as Sandbar and Blue sharks consume 0.2% to 0.6% of their body weight each day and take only three to four days to digest a meal. Very active sharks such as makos, which are warm-blooded (see "Are sharks cold-blooded or warm-blooded?" in chapter 2), consume 3% of their body weight per day, digesting a meal in 1.5 to 2.0 days. A 100-kg (220-lb) mako would therefore have to eat 3 kg (6.6 lb) of fish per day. It appears that sharks digest most of one meal before they resume feeding. And as high as these numbers may seem, they are only about half what a comparable-sized bony fish would need to maintain itself.

Low energy demands appear to characterize even the superpredatory White Shark. A 4.3-m-long (14-ft) White Shark was outfitted (harpooned actually) with a transmitting thermometer. Researchers were able to calculate its oxygen and energy demands by looking at how quickly its internal temperature changed as it swam through water masses of differing temperatures. The shark used up energy at an hourly rate of 0.2 calories for each kilogram of body weight (0.2 Kcal/kg/hr). By comparison, bony fishes use about twice that amount, and humans use up about eight times as much energy per unit body weight.

We can translate these numbers into how much a White Shark would need to eat to maintain itself. White sharks often feed on dead whales. Whale blubber has a very high caloric value: a half pound of whale blubber (the size of a tuna fish can) provides 1,600 calories. Researchers captured a shark that was feeding on a dead whale and found 30 kg (66 lb) of whale blubber in the shark's stomach. At an energy consumption rate of 0.2 Kcal/kg/hr, a belly full of whale blubber could maintain a White Shark for about

six weeks, which is plenty of time to find another dead whale. Energy-wise, sharks live in the slow lane.

Do sharks chew their food?

We have to chew our food to make it soft enough and especially small enough to swallow. The same holds for most animals that can't bite off chunks of food. They are "gape-limited," meaning they are restricted to eating small, swallowable things. Sharks are among the few fishes that are not gape-limited because they have the sharp, cutting teeth and strong, mobile jaws that allow them to tear and cut their prey into sizes they can swallow. All this is accomplished with the teeth that line the outer jaws, or in the case of sawfishes and saw sharks, with the teeth along the rostral "saw".

So most sharks don't chew their food, except for the groups discussed in the next question. Plus one other. Juvenile Lesser Spotted Dogfish create swallowable pieces of prey in an entirely unique manner. They anchor large food items on the pointed scales (dermal denticles) on the top of their tail, grab the food in their jaws, and pull, thereby tearing off bite-sized pieces.

How do sharks eat hard-shelled animals?

Food must be small enough and soft enough to swallow and digest. Some sharks, skates, rays, and chimaeras feed on hard-bodied prey, animals that have evolved tough outer hides and shells that protect the soft and nutritious stuff inside. Such food types have to be opened or crushed before they become edible.

Animals that feed on hard-shelled prey are called durophages, *phage* meaning "to eat" and *duro* meaning "tough" or "long-lasting," so literally, "eaters of tough prey" (from the same roots as "durable," "enduring," and the battery trademark "Duracell"). Durophagous species have specialized teeth for crushing the shells of the sea urchins, mollusks (clams, snails, etc.), and crustaceans (crabs, lobsters) that make up their diet (see "What kind of teeth do sharks have?" in chapter 2). Durophages, unlike more usual elasmobranchs, have one of three adaptations to crush their prey: (1) differently shaped teeth in the front versus the sides of each jaw (Port Jackson Sharks, Horn Sharks, Bonnethead Sharks); or (2) multiple rows of flattened teeth in their jaws which create broad surfaces in both jaws that can be brought together to crush prey (Nurse Sharks, many rays); or (3) fused, parrotlike beaks (holocephalans).

It is not only the teeth that set durophagous elasmobranchs apart. The

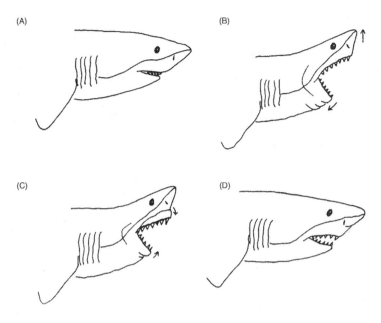

The sequence of head and jaw movements that go into the bite of a White Shark. (A) Normal resting position; (B) snout is lifted and lower jaw is dropped, achieving maximum mouth opening; (C) lower jaw is lifted forward and upward, and upper jaw is rotated forward and downward, closing the jaws—the actions that constitute the bite; (D) snout is dropped back down and upper jaw is pulled back to resting position. The bite (component C) occurs quickly, in about 0.8 second. From T. C. Tricas and J. E. McCosker, "Predatory behavior of the White Shark (*Carcharodon carcharias*), with notes on its biology," *Proceedings of the California Academy of Sciences* 43 (1984): 221–38; used with permission

strength of the bite is greater in durophages than in feeders on softer-bodied prey. Horn Sharks and Spotted Ratfish have higher bite forces for their body size than do fish- and squid-eating sharks such as Velvet Belly Lantern Sharks and Spiny Dogfish. Cownose Rays generate bite forces 20 times greater than what is needed to crush the clams they eat. Such force could damage the jaws if it weren't for special adaptations in the jawbones. The myliobatid Bat and Cownose rays have calcified struts in their jawbones lined up with the crushing plates. These hardened struts keep their jaws from twisting or collapsing when they are biting down on hard items. Bull Sharks and White Sharks have the strongest bite force at any given body size, probably because they feed on tough-bodied marine mammals and turtles. Larger jaw muscles, increased leverage between upper and lower jaws, dentition specialized for crushing, and mineralized jawbones and struts—taken together, these features provide durophagous chondrichthyans with the feeding equipment necessary for handling prey defended by toughness.

Do sharks store their food?

It is highly unlikely that sharks store food any place other than in their stomachs. Water is a poor medium for food storage because it carries odors so well. This makes it nearly impossible to hide a food item from scavengers. It's likely that there are other sharks around who, with their extremely

Nurse Sharks feed on a variety of hard-bodied prey and have teeth specially adapted for crushing rather than biting. Each tooth is flattened and has a rounded tip, which means that when the jaws are brought together, the teeth grasp and crush at the same time. Photo courtesy of Phil Motta

acute sense of smell, would find it, not to mention a host of scavenging crabs, bristle stars, snails, and other fishes.

Must a shark turn on its back to feed?

No. Although some sharks will turn on their sides or even their backs to feed, they don't have to. Sharks approach food at the surface and underwater head-on, swim over it, and then bite. Because their snout is slightly hinged and their upper jaw can be protruded (stuck out), sharks lift their snout to bite at things directly in front of them. We can't do that; we can only drop our lower jaw when we bite something.

Do sharks cooperate with one another to feed?

Cooperative feeding requires both intelligence and sociality. In killer whales, humpback whales, lions, and wolves, individuals do coordinate their movements, with different individuals playing different roles. Additionally, individuals must show restraint during the initial stages of attack. Otherwise, prey recognize the intent of the predators and invariably escape.

Cooperative feeding in bony fishes is rare, limited to Sailfish, Bluefin Tuna, pompanos, and lionfish. Getting into the water to watch sailfish feed actively is exciting and a little scary. Exposing oneself to actively feeding sharks is downright dangerous and observations are few.

At least three species of sharks have been observed feeding cooperatively. The best known of these, for which repeated observations exist, is the Broadnose Sevengill Shark. Sevengills are solitary predators on smaller prey such as fishes and other sharks. But when attacking large prey capable of defending themselves, such as Cape fur seals (*Arctocephalus pusillus pusillus*), which can weigh 350 kg (770 lb), they attack in groups. In a typical attack sequence, a dozen or so Sevengills approach a fur seal at the surface,

Broadnose Sevengill Sharks feed cooperatively on fur seals. (*a*) The sharks first form a loose circle around the prey and (*b*) then tighten the circle and prevent any attempts at escape; (*c*) eventually one shark launches an attack, at which time (*d*) the rest of the group shares in the kill. From D. A. Ebert, "Observations on the predatory behaviour of the Sevengill Shark *Notorhynchus cepedianus*," *South African Journal of Marine Science* 11 (1991): 455–65; reproduced with permission, © NISC (Pty) Ltd

surround it, and progressively tighten the circle. Escape attempts by the seal are thwarted by individual sharks that refrain from attacking it but force it back to the center of the tightening circle. Eventually, one or more Sevengills actually bite the seal, which stimulates attacks by the rest of the pack. Sevengills are also thought to use cooperative feeding to eat other large prey such as large Bat Rays, Spotted Gully Sharks (*Triakis megalopterus*), porpoises, and other Sevengill Sharks. All of the elements of cooperative feeding were observed in the reported attacks, including coordinated movement, division of labor, and initial restraint.

Other sharks that are thought to feed cooperatively include Sand Tigers off North Carolina and Australia, and Blacktip Reef Sharks in the Indo-Pacific. Sand Tigers have been observed herding prey, usually schooling fishes. In the waters off North Carolina, a school of 100 or more Sand Tigers, a species that is usually solitary, surrounded a school of Bluefish (*Pomatomus saltatrix*) and herded them into very shallow water, shallower than Bluefish and Sand Tigers normally occupy. The sharks then launched an all-out attack on the nearly stranded Bluefish. A group of Sand Tigers off New South Wales, Australia (where they are called Gray Nurse Sharks), herded a small shoal of juvenile Yellowtail Kingfish (*Seriola lalandi*) into a tight group before feeding on them. These sharks herded the prey by whipping their tails and creating sharp pressure waves that sounded like gunshots. The tail whipping may also have stunned the prey fish; killer whales also use this technique to feed on herring in the waters near Iceland.

This tail whip-cracking has been reported as a defensive behavior by the same shark in South Africa when cornered by divers in a cave (see "Are any sharks territorial?" in chapter 4).

Blacktip Reef Sharks near Palau, Western Caroline Islands, have been observed at least twice feeding perhaps cooperatively. Three Blacktips swam along a shoreline, "escorting" a school of 15 to 20 Milkfish (*Chanos chanos*). The sharks appeared to be keeping the Milkfish from moving into very shallow water by cutting off their access to the shoreline, while forcing them into a tight group. Different sharks intercepted the school if it moved into shallower water. The sharks' movement was slow except when individuals cut off the compacted Milkfish school if it moved toward the shore. No attacks occurred. The Milkfish school then swam into deeper water with the sharks following closely. On a second occasion, the same observer watched seven juvenile Blacktips that kept a group of about 20 small jacks (*Caranx sexfasciatus*) bottled up against the roots of a small mangrove islet. The Blacktips formed a semicircle, individuals occasionally darting towards the mangrove roots to attack any jacks that strayed beyond the mangrove plants. The observer speared one of the sharks and found a partially eaten jack in its stomach.

Other species of sharks (Oceanic Whitetip, Whitetip Reef, threshers) have been observed and reported as possible cooperative feeders, but the observations remain unconfirmed. Recognizing that sharks have relatively large brains, and that our knowledge of the catalogue of their behaviors grows in complexity with the years, it won't be surprising to learn that other species feed in a coordinated, cooperative manner.

Sharks in the family Carcharhinidae have an eyelid that moves up to cover the eye and protect it from physical harm. In this Caribbean Reef Shark (*Carcharhinus perezi*) this "nictating membrane" is half-closed. The mouth is to the left. Photo courtesy of Jeff Rotman, www.jeffrotman.com

Left eyes of two Common Stingarees, *Trygonoptera testacea*, a shallow-water stingray from Australia that has a retractable eyelid. The photo on the left shows the eye condition under relatively low light levels with the pupil cover retracted. On the right, the reflective cover is hanging down over the black pupil. The cover is thought to control the amount of sunlight entering the eye during the day. The opening to the right of the eye is the spiracle, through which water is drawn for breathing. Photos used with kind permission of Dave Harasti, www.daveharasti.com

A free-swimming, newborn Spiny Dogfish with its yolk sack still attached. Normally, shark pups use up their yolk supply and absorb the sack before birth, but this newborn was probably born prematurely when its mother was caught in a net. Photo courtesy of Jeff Rotman, www.jeffrotman.com

Blue Sharks produce large numbers of live young. This female contained 46 fully developed pups, a relatively small number for this species. Photo courtesy of H. L. Pratt

Southern Stingrays are attracted in large numbers to Stingray City, an area in the North Sound of Grand Cayman Island. Here, the water depth is less than 2 m (6.5 ft) and accessible to both snorkelers and scuba divers. Divers can feed and even pet the rays, which have become more than accustomed to people in the water. Photo by Kfulgham84, Wikimedia Commons, http://en.wikipedia.org/wiki/File:Stingray_CIty_in_Grand_Cayman.jpg

Most human injuries caused by sharks and their relatives result from provoked attacks involving a shark or ray defending itself against careless handling by a person. (A) A researcher studying deepwater sharks near the Azore Islands was handling a 110-cm (3.6-ft) Shortnosed Velvet Dogfish (*Centrosymnus cryptacanthus*) when it began thrashing around. The researcher reported that its "denticles ripped through my rubber gloves and left me bleeding on the deck…nothing velvet about it!" (B) A student in a fish biology class missed the lecture on the safe way to pick up a small stingray (by the spiracles) and suffered as a result. (C) Two snorkelers in shallow water in the Florida Keys encountered a small Nurse Shark lying on the bottom. One diver picked it up and threw it at his buddy, at which time the shark latched on to the second diver's thigh. *A,* photo courtesy of Maria Pickering; *B,* photo by Gene Helfman; *C,* photo courtesy of Adam Doss

The surfboard ridden by Bethany Hamilton when she was attacked by a Tiger Shark in Hawaii in 2003. The ruler is 1 ft (30 cm) long. The circumference of the bite on the board was 645 mm. Entering this number into an equation that calculates shark length from bite circumference for Tiger Sharks gives a minimum shark size of about 4.5 m (14.8 ft).

Photo by Randy Honebrink; used with permission of Randy Honebrink and Dayv Lowry

Cutting up a 10-m (33-ft), 8-ton Whale Shark with a chain saw. Whale Shark meat and fins are highly regarded in many Asian countries, where the animal is called Tofu Shark. Whale Shark meat has sold for as much US$15/kg ($7/lb), with fins from a single animal bringing US$400–$500. Although Whale Sharks are protected in many countries and their international trade is restricted, fisheries for the animal still exist in many locales. Photo © Wenn; source unknown

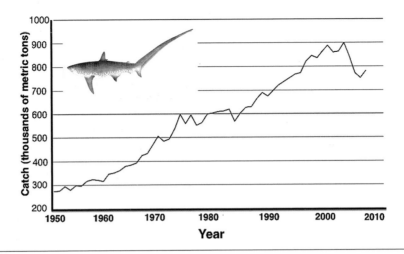

The number of sharks, skates, and rays caught in commercial fisheries has steadily increased in the past 50 years as shark products (meat, fins) have become more desirable. The reported numbers may represent only half of what is actually killed. The reproductive rate of most sharks is insufficient to keep up with the increase, with the result that many species appear to be in decline. Data from U.N. Food and Agricultural Organization; original graph by Chris Huh, http://en.wikipedia .org/wiki/File:Global_shark_catch_graph.png

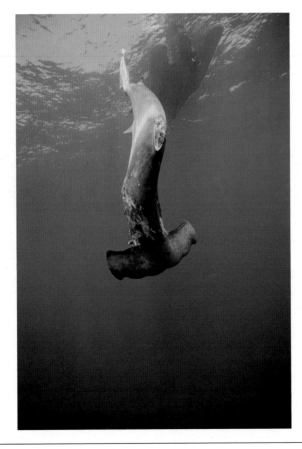

A 2-m (6.5-ft) Scalloped Hammer-head sinks slowly to the bottom off Cocos Island, Costa Rica. Its fins have been cut off, and the shark has been tossed overboard still alive, but it cannot survive or defend itself for long. Photo courtesy of Jeff Rotman, www .jeffrotman.com

A fourteenth-century amulet in the British Museum, London, made from a fossilized shark tooth set in silver, probably from a Megatooth Shark. Such pieces often had a silver loop at the top through which a thong or necklace could be strung. Called *glossopetrae*, ("tongue stones"), such fossils were thought to be the petrified tongues of dragons and snakes and were used as remedies against poisons and toxins. This amulet was about 3 cm (1.25 in) high. Photo by Gene Helfman

Contemporary native shark art. (*Left*) A "mola" from the San Blas Islands of Panama. Molas are sewn in reverse appliqué (multiple-layer) fashion onto the front and back of the blouses of Kuna women. The designs originated as body paintings but were incorporated into clothing following contact with Europeans. Many are now produced for the tourist trade. (*Right*) Detail from a Coast Salish house post, part of a sculpture titled "Interaction" by Musqueam artist Susan Point. The sculpture sits on the waterfront of Friday Harbor, San Juan Island, Washington. The post is carved and painted in the style of a totem pole, with a Spiny Dogfish near its base. Further up, the post includes herring, salmon, and, at the top, a killer whale, representing a healthy marine ecosystem. Photos by Gene Helfman

A Palauan storyboard. Storyboards depicted legends, with the action being shown from left to right in a single scene rather than in the multiple panels of Western cartoons. This storyboard tells of a fisherman whose canoe was bitten by a shark, shown swimming upside down under the canoe. Knowing that such behavior means that his wife is cheating on him, the fisherman returns to the village and kills his wife's lover. Photo by Gene Helfman

Stop in at any seaside curio shop, especially in south Florida, and you will find an array of shark goodies. We couldn't resist including a couple of local license plates seen while shopping. Photos by Gene Helfman

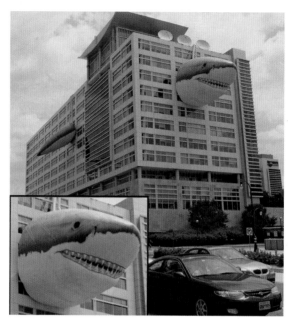

The Discovery Channel announced Shark Week each summer by decorating its headquarters in Silver Spring, Maryland, with a giant inflated White Shark. Head, tail, and pectoral fins stuck out from the building's back, front, and sides. The head is about three stories high. Despite its size, it is apparently modeled after a juvenile White Shark because the upper as well as the lower jaws contain the slender teeth characteristic of fish-eating juveniles (**see inset**). Photos by Gene Helfman

Winslow Homer's *The Gulf Stream*, which hangs in the Metropolitan Museum of Art, New York. Wikimedia Commons, http://en.wikipedia.org/wiki/File:Winslow_Homer_004.jpg

Chapter 8

Sharks and Humans

Do sharks make good pets?

This question actually has two parts: (1) Can sharks be kept in home aquariums? and (2) Do they make good pets? First off, note that the freshwater aquarium fishes called "sharks" are actually minnows, not sharks. Examples are Bala, Redtail, Variegated, Rainbow, and Redfin. They're lovely, easy to keep, small, and relatively cheap. They get their sharky names because of their body shape and a large, shark-like dorsal fin. But don't be fooled.

A number of actual sharks can be kept in home aquariums, although most eventually outgrow all but the largest living room tanks. Epaulette Sharks, Zebra Sharks, Bamboo and Brown-spotted Cat Sharks, Port Jackson Sharks, and Nurse Sharks are among the species offered for sale from aquarium suppliers, but almost all grow to be more than 1 m (3 ft) in length, are expensive, have stringent water quality requirements, and are mostly active at night (so not very interesting). For example,

- Epaulette Sharks ($350) are relatively sluggish bottom dwellers. They are attractive because they move across the bottom by "walking" on their pectoral and pelvic fins (but usually at night; by day, they hide in crevices). They will breed in captivity, but only in an aquarium larger than 510 L (135 gal). Advice on a website that sells them: "Care Level: Expert Only!!! Behavior: Aggressive. Minimum Tank Size: 180 Gallon."
- Zebra Sharks ($150), which are cute and colorful as pups, carry similar warnings: "Expert Only!!! Aggressive. Minimum Tank Size: 100 Gallon."

- Bamboo Sharks ($100): "Temperatures between 72 and 78 degrees Fahrenheit, pH of 8.1 or 8.4, specific gravity of 1.020 to 1.025. . . . Strong shark can easy dislodge unstable rock formations and corals. . . . All items in the tank should be stable."
- Port Jackson Sharks ($750): "Have a spine at the front of both dorsal fins that can cause injury . . . is a difficult fish to acclimate and may not eat. . . . Once eating well, it will eat virtually any crustacean or mollusk in the aquarium and will also prey upon any sleeping fish small enough for it to eat whole."
- Nurse sharks ($300): "Can reach a length of 4.3 m (14 ft) and a weight of 330 lbs (150 kg)." Let the buyer beware.

In addition to outgrowing their confined quarters, all elasmobranchs are carnivores, and most are predatory, which guarantees frustration if you want to keep more than one fish in a tank. Any shark can inflict a painful, damaging bite when handled.

Some saltwater rays are sold in the aquarium trade, but most grow too large for a home hobbyist. We have seen electric torpedo rays (Torpedinidae) for sale for as little as $80, but we can't imagine a more unsuitable and dangerous aquarium species. Freshwater stingrays (Potamotrygonidae) from the Amazon River basin are sold in pet stores and online. They have an advantage over sharks because they are freshwater animals and generally easier to keep than saltwater fishes. Some are bred in captivity, which is better than removing individuals from the wild. But they are expensive; a typical 8-inch-diameter (20-cm) Motoro Stingray sells for $200 to $300 or more (4-in stingrays may cost "only" $70)! And all but a couple of species grow to over 3 feet across, which makes them unsuitable for most home aquariums. Add to these problems the complication that it is illegal to sell or own a freshwater stingray in many states. As you can discover by reading posts on aquarium blogs, their care is complicated, their water quality requirements are exacting, and they go from apparently healthy to dead in a matter of hours (or shorter). And, oh yes, they do have that lethal barb, which gets fouled up in any kind of net if you have to move one for any reason, and then it can stab you as you try to untangle it.

Keeping freshwater fishes is relatively easy, but marine fishes—which means 99% of elasmobranchs—are another matter altogether. Marine fishes are more sensitive to water quality, temperature, food type, and disease than most freshwater fishes. Marine species are difficult to breed in captivity and thus must be captured in the wild, often with methods that destroy habitats and other animals. A responsible aquarist can avoid many problems by purchasing fishes that are on the American Marinelife Dealers Association's Ecolist of species that are easily maintained (see www.amdareef .com). Unfortunately, AMDA hasn't posted new information on its website

An Ocellate River Stingray at the New England Aquarium, Boston. These "Motoros" are one of the freshwater stingrays sold to home aquarists, although they can easily outgrow a living room tank. They are more appropriate for large public aquariums. Photo by Steven G. Johnson, Wikimedia Commons, en.wikipedia.org /wiki/File:Ocellate_river_stingray,_Boston _Aquarium.jpg

since 2007, and the organization may no longer exist, although its Ecolist can still be found. Regardless, fishes should be purchased from dealers certified by the Marine Aquarium Council (www.aquariumcouncil.org).

From the comments above, it should be clear that we are unenthusiastic about sharks, skates, rays, or chimaeras as pets for home hobbyists. We are not alone here. Even websites promoting home aquariums make statements such as "Most aquarists do best in avoiding these fish altogether." If you're looking for pet fish, there are plenty of alternatives to elasmobranchs among the many lovely, easy-to-keep, sustainably captured or cultured fishes sold by reputable dealers. Go with them and save yourself the heartache and the shark its life. Or buy a pair of Nike Air Jordan Grey Nurse Shark basketball shoes for $80. They're cheaper and will last longer.

What is the best way to take care of a pet shark?

Assuming you are ignoring our advice in the first question, read, read, read. Every shark has special issues related to temperature, tank size, water chemistry, food, lighting, structure, bottom type (many injure themselves on coarse rocks), tankmates, and diseases. To expect any success, you will have to educate yourself about the specific requirements of your shark. Websites that sell these animals often contain initial, enthusiastic statements ("one of the easiest sharks to keep in captivity . . . "), but if you do further research, you will also learn that mountain lions are one of the easiest large cats to keep in captivity.

The two most likely outcomes of owning a pet shark are (1) it will die, despite your large investment in time and money, or (2) it will live and outgrow your aquarium. Then what do you do? *Don't even think about releasing it into the wild*, where it will die, or transmit diseases, or eat native fishes. Leave shark keeping to the experts. Go to a public aquarium where paid

professionals and veterinarians work around the clock to give you an enjoyable, educational, environmentally friendly shark experience.

Are any medicines made from sharks?

Sharks have been and continue to be a direct source of some important medicines used to treat human ailments, although some of these medicines are more effective than others. Shark body parts used include cartilage tissues from skeleton and fin supports, oil and a chemical called squalene from shark livers, eyeballs (especially corneas), and various internal organs. Ancient Greeks and Romans used the electric discharges of torpedo rays to treat a variety of ailments. Scribonius Largus, the court physician to the Roman emperor Claudius (who adopted Nero), recorded using torpedo ray discharges to treat headache and gout. The ancient Greeks used such shock therapy as an anesthetic during operations and childbirth, the electricity functioning to numb the pain. The word "narcotic" has its roots in the Greek name for torpedo rays, *narke*.

Today, sharks are killed for chondroitin, a structural part of cartilage that is used as a dietary supplement for treating joint problems such as osteoarthritis. However, chondroitin can come from many sources, including from cows and pigs that are already at a slaughterhouse. It can also be made synthetically in laboratories. Sharks don't need to be killed to get chondroitin.

Some specialized types of eye surgery use chondroitin sulfate extracted from Blue Shark fins as a sort of glue that binds tissues together during cataract extractions, glaucoma procedures, and corneal transplants. The cartilage-derived "viscoelastic" has the right stickiness properties to work under these conditions. The manufacturers admit they could also use chondroitin sulfate from cow tracheas but don't because of concerns about transmitting mad cow disease. Another use of shark chondroitin sulfate from sharks is in skin grafts for burn victims. The chondroitin is combined with collagen from cow tendons and formed into an artificial "shark skin" called Integra.

Shark cartilage has also been promoted as a cancer cure. Because cancer in sharks is supposedly rare, shark cartilage was thought to prevent cancer in people. But sharks do get cancer, including cartilage cancers as well as melanomas and cutaneous fibrosarcomas (skin and connective tissue cancers). The U.S. Food and Drug Administration (FDA) and the Federal Trade Commission (FTC) filed injunctions against companies that made anticancer claims for shark cartilage products because clinical trials indicated that cartilage pills have little effect on cancerous cells.

Until recently, elasmobranchs were not used extensively in traditional Asian medicines, unlike bears (for their gall bladders) and tigers (testicles).

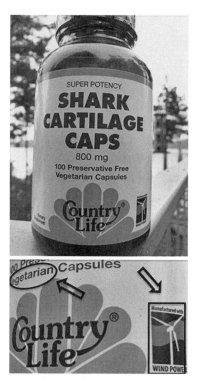

(*Top*) Untold numbers of sharks are killed each year to extract chondroitin from their cartilage to make cartilage pills, a product that can be made from many other, less threatened sources. (*Bottom*) But you can feel good because the outer capsule is made from vegetable matter, and the entire process is eco-friendly because it's wind-powered. Photo by Gene Helfman

One exception is sawfish snouts, which are employed in Chinese traditional medicine as a cure for heart disease, infections, parasites, weakness, and blood thinning. This use has fueled an unfortunate modern trade in sawfishes despite their endangered status and international protection.

Another alarming and recent exception involves manta rays and their relatives, which have found their way into Hong Kong markets in increasing numbers because of a demand for their gill plates. Mantas filter zooplankton out of the water with their gill plates (see the discussion of filter feeders in "How do sharks catch their prey?" in chapter 7). Proponents of manta medicine claim that eating gill plate soup, called "peng yu sai," can aid in the treatment of asthma, skin rashes, chicken pox, and, again, cancer. The erroneous logic applied is that since mantas filter food from the water, gill plate soup can filter and detoxify diseases from a person's body. There is absolutely no evidence to suggest that any of these benefits are real. Sadly, whereas only a limited market existed for mantas before (the meat is eaten in a few areas), they are now hunted extensively and in large numbers to feed the gill plate trade.

Sharks do produce one substance used in anticancer drugs. The active chemical is called squalamine and was first isolated from shark livers, stomachs, and gall bladders, not from cartilage. Squalamine has proven effective

in treating lung and ovarian cancers. Recent tests indicate squalamine may also be effective against various virus-caused diseases; if this turns out to be true, that would be a huge breakthrough in medicine because viral diseases are notoriously difficult to treat. Squalamine is patented and produced by reputable drug manufacturers. Most importantly, squalamine can be produced synthetically, so there's no longer a need to kill sharks to obtain it (see "Do sharks have diseases and are they contagious?" in chapter 9).

Before the 1950s, when it became possible to synthesize vitamin A, shark liver oil was a major source of vitamin A as a dietary supplement. "Shark liver oil is known to help strengthen and regenerate the immune system while benefiting many other functions and organs of the body," or so read the claims of companies that sold the product. Although shark liver oil has been used medicinally for centuries for ailments as diverse as wound healing, respiratory and digestive tract irritation, and lymph node swelling, none of these uses have been tested and proven medically, nor is shark liver oil prescribed or used by American medical doctors.

Some over-the-counter (nonprescription) medications contain shark liver oil. The most widely used of these is Preparation H for hemorrhoids. (Wikipedia has a graphically illustrated explanation of hemorrhoids—or ask any adult if you want the PG-13 version.) Shark liver oil is known to shrink blood vessels by restricting blood flow. Hemorrhoids result from inflamed blood vessels in the rectal region, so restricting blood flow to the area reduces the swelling and itching. Because of its effect on blood vessels, shark liver oil is also used in facial creams that target wrinkles.

Shark parts have proven useful in other medical procedures. One surprising application is using shark corneas in human corneal transplants. The cornea is the clear outer part of the eye in all vertebrates that light must pass through first before focused by the lens on the retina. Corneal disease causes the cornea to become cloudy, eventually blinding the victim. Shark corneas have a number of unique properties. Most importantly, because the collagen layers making up a shark's cornea are very dense, they don't swell in distilled water and remain transparent under a variety of surgical conditions. These traits have been known since the late 1800s, based on studies of the lowly Spiny Dogfish. Because shark corneas don't swell and human eyes don't reject shark corneal tissue, shark corneas are a superior tissue for transplanting into diseased human corneas. As a result, shark corneas have been used in corneal transplants for decades.

The usefulness of shark chondroitin sulfate for eye surgery and skin grafts, squalamine for cancer treatment, shark liver oil for treating various ailments, and shark corneas for transplants underscores another reason we should do everything possible to protect sharks. Sharks are a potential source of valuable products to cure human disease, including cures we have

yet to discover. But killing massive numbers of sharks for their cartilage as a source of chondroitin is avoidable, and as a cancer cure is foolish. Such unproven and unnecessary uses just promote the overexploitation of sharks.

Do sharks feel pain?

Our understanding of pain is highly colored by our own experience and reactions. When we talk about pain in other organisms, we must distinguish between whether they feel pain versus whether they feel pain they way we do.

Pain has a technical definition established by a scientific organization called the International Association for the Study of Pain (IASP). The IASP defines pain as "an unpleasant sensory or emotional experience associated with actual or potential tissue damage." The IASP goes on to list the elements of a sense of pain, including (1) appropriate anatomical, nervous, and physiological components; (2) behaviors that are more complicated than simple reflexes (such as pulling your hand away from the stove when you burn yourself); and (3) an ability to learn and therefore avoid future painful events.

If you want to know if sharks feel pain the same way that we do, the answer is probably no. Our sense of pain involves a part of our brain (the neocortex) that is lacking in sharks. This would suggest that sharks lack the first element in the IASP's list, namely, the appropriate structures. However, sharks do have pain receptors, called nociceptors, in their skin just like all vertebrates, so they can certainly sense painful stimuli. They also show internal stress responses similar to other vertebrates when caught on a hook and line—specifically, changes in blood chemistry (lowered blood pH, elevated cortisol levels). A number of studies show that sharks can learn as well as bony fishes, including learning to avoid nets after having been captured in one, which is certainly a stressful situation that could include painful stimuli. A growing body of experimental evidence strongly suggests that bony fishes feel pain by all the requirements of the IASP definition and criteria. Similar studies have yet to be performed on sharks, leaving the question ultimately up in the air.

The issue is additionally complicated by differences in the ability to ignore pain between different persons, and between people and sharks. Pain thresholds vary tremendously among people; women have a higher pain tolerance than men (confirmed in MythBusters episode 142, "No Pain, No Gain," which also confirmed that cursing out loud helps a person tolerate more pain). Sharks feed on stingrays, lionfish, and a variety of other spiny fishes that inject venom as part of their defensive adaptations. It may hurt to chomp down on a stingray—and some hammerhead sharks feed heavily

on stingrays—but how else are you going to get a meal? (Note that evolution doesn't select for elaborate traits that have no function. If specialized spines that injected venom didn't deter some predators via a painful, learnable reaction, the prey animals wouldn't waste all those genetic resources evolving them.)

Sharks engaged in one of their famous feeding frenzies will continue to feed even though they have been sliced open by another shark. In some cases, they will feed on their own digestive system that is spilling out of their wounds. Does this mean they don't feel pain, or is it that the excitement of the moment allows them to ignore the pain? (We've all experienced this to one degree or another during intense physical exertion.) We really don't know. Feeding frenzies occur when large amounts of food are suddenly available for a short period of time, just the sort of opportunity that sharks are built to take advantage of. Foregoing feeding would be a significant lost opportunity.

What should I do if I find an injured shark?

Sadly, there is probably little you can do to help an injured shark. It's probably best to leave it be. If it's a large shark, or if you know your sharks and think it's a rare species, a local university fish collection or public museum may be interested in the animal for research or display. Really big sharks are always of interest to the media, who will call an expert for identification, and you could wind up on the evening news (hopefully with a shark conservation message to contribute). Most importantly, realize that, as with any injured animal, handling it can be dangerous. Even small sharks can inflict painful and serious injuries, and wounded animals of all species are seldom as grateful for our efforts as we might like to believe.

What should I do if I find a shark that looks diseased?

As in the case of injured animals, it's best to leave a diseased shark alone. External lesions, sores, or other signs of disease may indicate water quality problems, such as pollution or harmful algal blooms. You could call a local health department, inform them of the location and particulars, and let them decide. Few things on the beach smell worse than a large decaying shark, so local officials will probably want to dispose of the animal quickly.

How can I see sharks in the wild?

Given the hundreds of sharks, skates, and rays that occur in shallow water around the world, it's likely that anytime you're in the ocean, there's

a shark nearby. That people don't see them more often attests to their ability to hide, or their tendency to flee, rather than to their absence. Because sharks usually don't want to be spotted, seeing them in the wild usually involves going to a place where they are known to aggregate and often involves luring them with food.

Shark encounter diving, or shark ecotourism, has grown tremendously in the past few decades. Many more people want to see sharks than catch them, and people are willing to spend large sums of money for the privilege. Since the same shark can only be caught once but can be watched over and over, economics clearly favor shark watching over shark fishing. Various studies have calculated how much a live shark is worth compared with one dead on the deck of a boat. In the Bahamas, a dead shark may be worth $40 to $50, whereas a live one is worth $750,000. In the Republic of Maldives, a dead Gray Reef Shark has a onetime value of $35 but an ecotourism value of $3,300 to $33,500 per year depending on locale. In Moorea, French Polynesia, Sicklefin Lemon Sharks (*Negaprion acutidens*) form the basis of ecotourism diving, with the lifetime value of each shark estimated at $2.6 million. And in Palau, Western Caroline Islands, where diving tourism is a mainstay of the national economy, a reef shark is worth $108 for its fins and meat but $1.9 million live and swimming around the reef. Not surprisingly, Palau has declared its entire territorial waters a shark sanctuary, outlawing the killing or capture of all sharks.

While you may be fortunate enough to encounter a shark anywhere you dive, a few places have capitalized on shark and ray abundance, seasonality, water clarity, and boat access:

- Established White Shark cage diving occurs near Isla Guadalupe in Baja California, Mexico; near Cape Town, South Africa; at the Farallon Islands off San Francisco; and off Neptune Islands, South Australia.
- Whale Shark encounters, estimated to be worth $47.5 million dollars annually, can be had at Ningaloo Reef, Western Australia; the Sea of Cortez and Yucatan Peninsula areas of Mexico; Belize and Honduras in the western Caribbean; Donsol Bay, Luzon, the Philippines; Sodwana Bay, Kwazulu-Natal, South Africa; and the Maldives.
- Viewing hammerhead sharks requires long-distance plane and boat travel, as they are usually in places and at depths accessible mostly to experienced divers. Where they occur, they can be found in spectacular numbers. Reliable spots include the Cocos Islands of Costa Rica; Green Island, Taiwan; the Wolf and Darwin islands in the northern Galapagos Islands; and Malpelo Island, 235 miles (378 km) west of Colombia, where they have been best studied and where large aggregations of Silky Sharks also occur.

A 3- to 3.5-m (10- to 11.5-ft) White Shark swims close by observers in a shark cage in Shark Alley, near Dyer Island, South Africa. Dyer Island is one of the premier locations for viewing White Sharks in the wild, one reason South Africa was among the first countries to outlaw fishing for the species. Photo by Gene Helfman

- Some of the better-known places to see a variety of nearshore and oceanic sharks include Grand Bahama Island for Caribbean Reef Sharks and occasional Tiger and Lemon Sharks; offshore of Oahu, Hawaii, for Sandbar Sharks and Galapagos Sharks; and Blue Corner, Peleliu Island, Palau, for Gray Reef Sharks. Some of these places offer cage dives, and some involve free dives. For coldwater enthusiasts, Elliott Bay near Seattle, Washington (just off the Seattle Aquarium pier), has a shallow-water population of Bluntnose Sixgill Sharks, a species that normally occurs at great depths.
- Manta Rays aggregate at known times and places off Yap and Palau, Western Caroline Islands; Manta Point, Bali, Indonesia; Tofo in southern Mozambique; Hanifaru Bay, the Maldives; and Kona, Hawaii.
- Southern Stingrays at Stingray City, Grand Cayman, Cayman Islands, have grown comfortable with divers and gather at predictable locations where people give them handouts. The rays tend to remain at the feeding site, creating aggregations of animals that are normally quite solitary. These rays have lower condition factors (body weight as a function of body length), higher external parasite loads, and more bites on their bodies due to fighting than nearby rays in areas where they aren't fed by tourists.

Shark ecotourism has its downsides. When shark fishing is banned in favor of shark watching, fishers may not benefit from the large profits associated with shark watching. International tourists are seldom willing to dive from the average fishing boat, and fishers can seldom afford the boats and dive gear, cages, food service, and the like expected in tourist diving operations. Another controversy surrounds interactive diving and in-water shark feeding, where dive leaders or even tourist divers hand-feed sharks attracted to certain areas by bait and chopped up fish parts ("chum"). A major concern is that sharks come to associate humans with food. Sharks near feeding sites now associate the low-frequency sound of a boat's motor

Sharks: The Animal Answer Guide

with free food, arriving well before any chum hits the water, with obvious implications to user groups not interested in having a sharky experience. Animals that learn to associate people with a free meal can become aggressive if you show up without a handout. An aggressive chipmunk is one matter; an aggressive reef shark is another.

To minimize such associative learning, White Shark cage dive operations in South Africa use chum to lure sharks in, and toss out a chunk of tuna on a rope to bring them even closer, but don't let the sharks eat the bait. At other locales, irresponsible dive operators have gone so far as to stuff old wetsuits with bait, which makes for dramatic video footage but is almost certainly a bad habit to encourage on the part of the sharks. The economic losses that could result if an attack occurs (and attacks have happened, including at least one fatality in the Bahamas) could negate all the positive economic benefits of the shark watching trade and spill over to nonshark tourism. Because of such concerns, the feeding of sharks by divers has been banned in several locales, including Florida, the Maldives, the Cayman Islands, and Hawaii.

How can I see sharks in captivity?

Just about every public aquarium of any size has a live shark exhibit. Some specialize in local species; others have tanks with both local and exotic (usually tropical) species. Among the large active species exhibited are Sand Tiger, Sandbar, Blacktip Reef, Leopard, and Spiny Dogfish sharks and Bat, Spotted Eagle, and Cownose rays. Sawfishes, Nurse Sharks, wobbegongs, and various skates and stingrays fill out the list of commonly displayed elasmobranchs. Sand Tigers are an especially popular large species because they live a long time in captivity, swim slowly above the bottom, and have multiple rows of exposed spiky teeth that draw much attention. A frequently published figure of 30,000 for the number of teeth a shark produces in its lifetime is based on systematic, daily sieving of the sand at the bottom of a tank in which Sand Tigers were kept.

Many public aquariums feature shallow "touch tanks" where visitors can get up close and very personal with small, relatively harmless species. Touch tanks allow viewers to contact animals in a friendly, supervised manner. Popular prisoners in these interactive exhibits are Bonnethead Sharks, Bat Rays, and Cownose Rays, the latter two with their barbed stings clipped. A touch tank with Cownose Rays is even available next to the right field stands at Tropicana Field in St. Petersburg, Florida, home of the Tampa Bay Rays baseball team.

Smaller tanks typically display species that also show up in the home aquarium trade. These are usually small (less than 1 m, or 3 ft, long) Epaulette

Sand Tiger Sharks get their alternative name, Ragged Tooth, because of their long, sharp teeth, which are especially suited for grabbing slippery prey such as fishes and squid. Photo by Gene Helfman

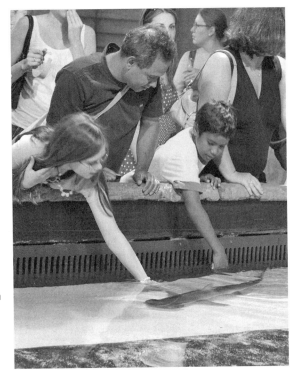

Kids of all ages crowd around the Touch Tank at the Georgia Aquarium in Atlanta. The tank contains small Bonnethead Sharks and Cownose Rays, the latter with their tail barbs snipped off. Photo by Gene Helfman

Sharks, Zebra Sharks, Bamboo or Brown-spotted Cat Sharks, Port Jackson Sharks, and their relatives. When public aquariums have chimaeras, they are almost always in a separate, smaller, darkened tank illuminated with red light because chimaeras avoid brightly lit habitats.

The Monterey Bay Aquarium in California has been the first public aquarium to successfully exhibit White Sharks, despite efforts at several other facilities. Since 2004, MBA has held a half dozen 4- to 5-foot-long

Sharks: The Animal Answer Guide

juvenile White Sharks in its huge Open Sea tank for periods ranging from 11 days to up to 6½ months before tagging and releasing the sharks back into the wild. Five of the six survived their releases and were tracked successfully for varying periods. (One died in a fisherman's net.)

The Georgia Aquarium in Atlanta is the only facility in the United States that has been able to display very large elasmobranchs; it has manta rays and Whale Sharks in its Ocean Voyager tank. The Georgia Aquarium came under strong criticism in 2007 when two of its much-publicized Whale Sharks, Ralph and Norton, died.

The Okinawa Churaumi Aquarium and the Lisbon Oceanarium have also kept manta rays successfully, and the Okinawa rays produced live offspring from two births. The Churaumi Aquarium has also kept Whale Sharks successfully, as have the Osaka Aquarium Kaiyukan and the Kagoshima Ioworld Aquarium in Japan, the Kenting National Museum of Biology and Aquarium in Taiwan, and the Polar Ocean World in Qingdao, China.

Chapter 9

Shark Problems (from a human's viewpoint)

Are some sharks pests?

Ask anglers in the Puget Sound region of Washington State, in coastal British Columbia, and in many other places along the U.S. Pacific Coast, and they will tell you that Spiny Dogfish are a real nuisance. If you troll a herring while fishing for salmon, or bottomfish with bait for rockfish or Lingcod, you are much more likely to catch a dogfish than a salmon or ling. Getting the dogfish off your hook or out of a landing net without being nailed by its long, sharp, venomous spines can be a tricky business. Nurse Sharks are similarly viewed as something of a nuisance in some parts of the Caribbean because of their bait-stealing habits as well as because of their tendency to destroy fish traps while trying to get at the reef fish inside.

Anglers in many other places consider sharks a nuisance because a fish struggling on a fishing line is like ringing a dinner bell for nearby sharks. In some places, it's almost impossible to land a tuna or a billfish without having it attacked and often eaten (except for the head) by a shark. (Read Hemingway's *The Old Man and the Sea* for a fascinating account.) Long tooth marks on shark-decapitated tuna indicate that the shark swallows the tuna whole from behind, then slips back and bites down just behind the gills, eating the fleshy body and leaving behind the bony head for the frustrated angler.

Can there be too many sharks in an area?

Some people feel that any number of large sharks in a place where people swim is too many. But given the millions of people that swim in the

A mountain of dogfish. Fishers targeting other species along the Pacific Coast of the United States and Canada can wind up with more Spiny Dogfish than anything else, as happened with this catch off southern California. Photo by John Wallace, www .photolib.noaa.gov/htmls/fish0221.htm

ocean and the extremely low number of attacks that occur each year, we feel such concerns are unjustified (see the next question).

Someone swimming or fishing might find having a few sharks around troubling. But sharks seldom if ever reach the abundance achieved by some other elasmobranchs whose numbers do have serious economic and ecological impacts. Feeding schools of Cownose Rays along the eastern U.S. seaboard, commonly in shallow bays such as the Chesapeake, can number in the hundreds. The rays move across grass, mud, and sand flats, exposing prey with their flapping pectoral fins and then suctioning up food organisms with their mouths. They crush shellfish such as clams, oysters, and scallops using their pavement dentition. Their large numbers and high rates of shellfish consumption have led to proposals to create a commercial fishery targeting rays as a means of reducing ray predation on such commercially important organisms as clams and oysters.

Are sharks dangerous to people?

Before 1916, there was little fear of sharks in North America simply because there had been no documented shark attacks on humans in temperate waters (ignoring reports from other temperate countries such as Australia and South Africa). In 1891, Hermann Oelrichs, a banker and adventurer, even put up a sizable reward for anyone who could document a shark attack in the temperate waters of North America. Historian Michael Capuzzo concluded that respected scientists, shark experts included, were of the opinion that "sharks were not capable of inflicting serious injury to man."

Everything changed in 1916, a story detailed in Capuzzo's *Close to Shore: The Terrifying Shark Attacks of 1916* and Richard Fernicola's *Twelve Days of Terror: A Definitive Investigation of the 1916 New Jersey Shark Attacks*. In July of that year, over a span of less than two weeks, four people along New Jersey shores and creeks were killed by a shark, or sharks. The shark or sharks involved were either White Sharks or Bull Sharks, and many eye-witnesses were present to confirm the perpetrators as sharks. The attacks ended, causally or coincidentally, when a 2.25-m (7.5-ft) White Shark was captured in the area. It contained human body parts. These incidents inspired Peter Benchley's book *Jaws*.

About 75 unprovoked shark attacks on humans occur each year worldwide. Of these, about five (7%) are fatal. Somewhat more than half these annual attacks, about 40, occur in the United States, with about one-third in Florida; other states where attacks typically occur include Hawaii, California, South and North Carolina, Texas, Oregon, New Jersey, and Georgia. All U.S. coastal states except Maryland have seen incidents. In the more than three centuries since 1670, only 44 fatalities have occurred in U.S. waters; over the past decade the United States has averaged less than one fatal attack a year.

These statistics come from the International Shark Attack File (ISAF), a catalogue managed by your second author (GB). ISAF is based at the Florida Museum of Natural History, at the University of Florida. ISAF has kept records of shark attacks for more than 50 years, including accounts from the mid-1500s. Staffers at ISAF receive reports and read news stories daily. All leads are referred to a network of biologists, doctors, and other informants stationed near coastlines around the world. A local representative makes contact with the victim of the attack—or the authorities, in the case of a fatal bite—and gathers as much information as possible. A standard Shark Attack File questionnaire is completed, with data on weather, tides, and how both the shark and the victim behaved before, during, and after the attack. Annual summaries are available at www.flmnh.ufl.edu/fish/sharks/isaf/isaf.htm.

Any shark can inflict a serious wound; these are, after all, predators that feed and defend themselves with their powerful teeth and jaws. ISAF has records of 27 different species involved in unprovoked attacks. But some sharks (and many rays) are involved in provoked attacks where people handle or step on them or feed them.

Positive identification of a shark species involved in an attack is uncommon and difficult. (Not surprisingly, most victims are too distracted to notice key identification characters during an attack.) In rare instances, a tooth fragment left in the victim or the surfboard facilitates identification. That said, three species have been identified as responsible for more than half of all attacks: White, Bull, and Tiger sharks.

This surfboard and its rider were attacked by a White Shark near Santa Barbara, California, in October 2012. The shark was estimated to be around 5 m (15–16 ft) long. The impression left by the shark's initial bite is obvious. Unfortunately, the victim did not survive. Photo by Santa Barbara County Sheriff's Department

All three species occur widely, are large, and frequent nearshore waters where people swim and recreate. All three have teeth especially equipped for shearing away flesh from the large animals they often eat. White Sharks attack a variety of marine mammals, from sea otters to whales. Bull Sharks eat sea lions, dolphins, sea turtles, and other sharks. Tiger Sharks eat sea turtles, other sharks, dolphins, and dugongs. To these sharks, we are undoubtedly just another potential prey item of the right size, one that swims slowly and awkwardly at that.

Bull Sharks, also known as Zambezi Sharks in South Africa, are cited as the second most common perpetrator of attacks on people after White Sharks, and the most common attacker in tropical and subtropical waters. Bull Sharks frequent nearshore areas and commonly enter rivers, the same places where people swim, bathe, and wash their clothes. Bull Sharks have been or are suspected to have been responsible for attacks in rivers in the United States, Australia, Iraq, Iran, India, and Nicaragua (in Lake Nicaragua and its outlet river), in a variety of marine habitats from reefs to beaches. Pound for pound this may be the most dangerous shark in the sea.

Tiger Sharks are another tropical species that will also enter water barely deep enough for them to swim. Surfer Bethany Hamilton was attacked by a Tiger Shark off the island of Kauai, surviving with the loss of an arm. Tigers are very active at night and are probably responsible for many attacks on people swimming in the dark in tropical waters.

White sharks attack people: of this there can be no doubt. But do they want to eat us? Surprisingly, the evidence strongly suggests that they don't. Most White Shark attacks on people (about 60%) involve a single bite, after which the shark leaves, suggesting an investigatory attack followed by rejection. But perhaps we are better than seals at getting out of the water before the beast returns for "seconds," its natural feeding behavior. Many attacks occur near seal and sea lion colonies, so it is possible that in such attacks the shark mistakes the swimmer, diver, or surfer for another species. Because White Sharks launch such attacks from below, the outline of the

backlit surfer or swimmer is not that different from that of a seal (see the discussion of ambushers in "How do sharks catch their prey?" in chapter 7).

White Sharks also attack divers who have speared a fish. Whether the shark was investigating the sounds of a struggling fish or smelled fish blood is unknown, but the human hunter then becomes prey. Abalone and sea urchin divers frequently encounter White Sharks and attacks do occur, perhaps because the divers are easy pickings as they hunch over their prey rather than looking over their shoulders for bigger predators. Few people collect abalone and sea urchins in waters where large White Sharks abound, but there are exceptions, and some have paid a price. One picker has been bitten three times by White Sharks in California waters but has kept on diving.

Surprisingly, the most dangerous shark in U.S. waters, in terms of number of people bitten, may not be the White, Bull, or Tiger (or Hammerhead, Lemon, Reef, Sand Tiger, mako, Blue, or White Tip). It is the Blacktip Shark, a moderate-sized (2-m, or 6- to 7-ft) fish eater that occurs commonly in the surf zone. Blacktip and Spinner (*Carcharhinus brevipinna*) sharks are probably responsible for three-quarters of U.S. attacks, about 30 a year. We say "probably" because the average victim is not able to identify those two species, and the tooth fragments that experts use for identification are seldom recovered. Victims are usually young males in their teens or early 20s who are surfing or boogie-boarding in the surf zone.

Another surprising perpetrator of many attacks is the Nurse Shark. The most common "victims" of Nurse Shark attacks are again young males in their teens or early 20s (do we see a pattern here?) who are snorkeling or scuba diving in shallow water. Nurse Sharks don't generally attack unless provoked, something that happens more than it should because some people can't resist the temptation of pulling the tail of a Nurse Shark lying on the bottom. Nurse sharks are among the few sharks flexible enough to put their tail in their mouth, so an attack is clearly the animal's defensive response to having its tail pulled. Divers have also been bitten petting Nurse Sharks, and a Florida diver in 2006 kissed one and received "lacerations to upper lip." Although Nurse Sharks have relatively small teeth, they can still do damage because they tend to bite and not release their grip. As a result, divers find a nurse shark attached to their hand or arm or chest or leg, one that requires forceful, drastic means to remove. Such provoked attacks are almost never fatal (except for the shark), although the round, craterlike wounds are prone to nasty infections and often require skin grafts.

In the category of bizarre, unprovoked attacks we return once again to the Cookiecutter. Before 2009, Cookiecutters did not appear in any of the various international "incident" or "attack" files. But in March of that year a swimmer attempting to cross the channel between the Big Island of Ha-

A Nurse Shark cruises slowly over the reef at Ambergris Caye, Belize. Nurse Sharks are often encountered lying perfectly still on the sea floor in shallow Caribbean habitats. They are docile, harmless sharks that can grow to over 4 m (13 ft) in length. "Attacks" are almost always provoked, as when a diver pulls the tail of a resting shark. Photo by Joseph Thomas, Wikimedia Commons, http://en.wikipedia.org/wiki/File:Nurse_Shark_4472.jpg

waii and Maui (at night) was attacked twice by a Cookiecutter Shark. The shark first tried biting the swimmer's chest and, not finding much meat there, removed a circular plug of flesh three inches across and one inch deep from his calf, a classic Cookiecutter mode of attack. The wound took nine months to heal completely.

Stingrays are obviously dangerous because of their toxin-covered barbs. Most stingray "attacks" are provoked, in the sense that the injured person deliberately or accidentally grabbed, kicked, or stepped on the animal. Totally unprovoked incidents are also known, however. A woman riding in a fast-moving speedboat in the Florida Keys in 2008 was killed when a 75-pound Spotted Eagle Ray jumped out of the water and landed on her. She struck her head against the boat and died from blunt force trauma. Eagle and manta rays frequently leap entirely out of the water for reasons that aren't altogether clear. One idea is that they are trying to rid themselves of attached remoras (suckerfish). Another Florida incident in 2006 also involved a Spotted Eagle Ray leaping out of the water and into a boat and, in the process, stabbing a man in the chest. Fortunately, the man survived.

Why do sharks attack people?

The International Shark Attack File classifies attacks as either unprovoked or provoked. The reasons for unprovoked attacks are varied. Among the suggested motives are (1) investigatory bumps and bites, where a shark is testing the edibility of an item; (2) defensive or aggressive bites, where a person is somehow perceived as a threat or as a competitor; (3) territorial defense, although true territoriality in sharks is not known; (4) mistaken

identity, where the shark mistakes a person for a more usual prey item; and (5) true feeding attacks.

Provoked attacks are the ones that occur when someone accidentally or deliberately disturbs a shark. These are clearly not instances of sharks trying to make a meal of someone but are more likely defensive reactions by the sharks. Other provoked attacks are the result of careless handling. Feeding and chumming for sharks also lead to provoked incidents, with an excited shark biting the feeder.

The placoid scales of many sharks are sharp and can cause injury, and people bumped by sharks are sometimes cut or scraped. Whether bumping is provoked or unprovoked—or should even be considered an attack—can be debated.

Unprovoked attacks by ray species are rare (unless you're a clam). Provoked attacks, however, are all too common and sometimes fatal. Electric (torpedo) rays (Narcinidae, Torpedinidae) defend themselves by discharging powerful electric shocks. Fishers who handle a torpedo ray are invariably shocked, and the discharge is strong enough to stun a person or worse. We've seen grown men knocked to the deck of the boat from the discharge. Some scuba drownings in southern California may have been caused by encounters with Pacific Electric Rays (see "How do sharks detect electric fields?" in chapter 2).

Stingray attacks are also invariably defensive. A wader steps on a ray, and the ray responds by flexing its tail upward and forward, driving its spine into the wader's foot or lower leg. This spine has serrated sawtooth edges coated with a protein-based venom and mucous that diffuses rapidly and breaks down the victim's tissues. Stingray stings cause immediate, intense pain, at least in people. The sawtooth edges create a large wound when the ray or the person pulls back or removes the spine. Victims of stingray wounds often describe it as the most painful thing they've ever experienced.

Fatal incidents involving stingrays are fortunately rare (no one keeps a record of the number, unlike shark attacks). Sometimes divers mistake large dasyatid stingrays (with spines) that are lying on the bottom for manta rays, which don't have spines—and don't lie on the bottom. A diver may try to take a ride, grabbing onto the front edges of the ray's disk. The ray again responds defensively, driving its tail and spine upward and forward, but this time into the diver's chest region. Fatalities from such incidents have occurred. A documented death occurred in Sulawesi, Indonesia, when a diver tried to hitch a ride on a large stingray. Speculation has it that "Crocodile Hunter" Steve Irwin, who was killed by a large ray while diving on the Great Barrier Reef in 2006, swam over or close to the ray, and the ray responded with a defensive tail thrust.

A Pacific Electric Ray, an animal suspected of causing drownings of divers who bump into it accidentally at night and receive a debilitating shock. Photo by Daniel Gotshall, Wikimedia Commons, http://en.wikipedia.org/wiki/File:Torpedoray_300.jpg

How likely are shark attacks?

Being attacked by a shark is highly unlikely. Unpleasant, but still unlikely. When you consider the millions of people who swim, work, and play in waters frequented by dangerous sharks, the number of attacks is minuscule. You are much more likely to win the lottery than be attacked, and your chances of winning the lottery are one in several million. (Actually, your chance of a $1 million payout in Powerball is 1 in more than 5 million; see www.powerball.com/powerball/pb_prizes.asp.) Your odds of being attacked are 1 in 11 million, and of dying, less than 1 in 264 million. Put in another context, half the U.S. population, about 167 million people, lives in coastal counties. That's about one-half death per year for the entire U.S. coastline. Pretty good odds.

Many other activities are considerably more dangerous than swimming around sharks. In the United States alone, three dozen or so people are attacked by sharks annually, with maybe one death. In contrast, many more people die from toaster accidents (17 in the United States in 2007); from suffocating in collapsing sand holes while playing on the beach; or from stabbing and trampling injuries from deer, jellyfish stings, roller coaster accidents, and ant bites and stings. Alligator attacks are two to three times more likely to be fatal. Almost four times as many people are killed by falling vending machines as by sharks, but these may again be provoked attacks: most such tip-over incidents involve people who are enraged by jammed candy bars or who try to shake a freebie out of the machine. Rocking a 500-pound vending machine back and forth to get a free bag of Cheetos may qualify you for a Darwin Award (see www.DarwinAwards.com). Going for a swim at the beach is comparatively safe and sane.

Can scientists estimate the size and species of a shark from bite marks?

News accounts of shark attacks almost always speculate on what kind of shark was responsible and, of course, how big it was. Any attack—whether on a surfboard, a person, or a submarine—happens quickly (remember: most sharks rely on surprise to be successful). Consequently, victims or bystanders have only a fleeting glimpse of the animal involved. Even trained observers find making accurate assessments of size and species challenging.

This is where forensic analysis—a sort of CSI (crime shark investigation)—can contribute information. Research has shown that the size and even species involved can be estimated from as little as a few scratches on a surface produced by a shark's teeth. An investigator looks at any tooth fragments left behind, measures bite profiles, analyzes location and habitat, and compares the information with species distribution guides and museum collections of teeth and jaws from known sharks. Or the investigator calculates shark size from equations developed by careful measurements that compare tooth and jaw characteristics of sharks of known sizes.

You might think that the size of a tooth left behind would indicate shark size. But teeth vary between individuals and between locations in the jaw, as well as among the rows of teeth in the jaw, all reducing the accuracy of a size estimate. What about the width of the chunk taken out of a surfboard or of a wound on a whale carcass? Aren't these good indicators of shark size? Yes and no. The shark might not have gotten its entire mouth around an object, in which case we might underestimate the shark's size. Or the shark might have enlarged the wound by shaking its head or moving its jaws from side to side; this larger wound would suggest a larger shark than was actually the case. For this reason, analysis based on bite wound size will state that the shark was at least or about a certain size. This kind of reasoning was used in determining that the shark that attacked 13-year-old surfer Bethany Hamilton in Hawaii was a Tiger Shark with "a minimum length estimate of 4493 mm," given the circumference of the piece taken out of her surfboard.

Recent data indicate that the best forensic analysis of size comes not from the width of the bite but from measuring the distance between the impressions or holes left by the tips of the teeth on either side of the middle of the jaws, the "interdental distance," or IDD. The IDD is fairly constant for a particular size and species of shark, and even differs slightly between upper and lower jaws. Because a bite is likely to leave several tooth tip impressions above and below the bitten object, an investigator can make several measurements, and the more measurements that are made, the greater

Sharks: The Animal Answer Guide

the accuracy. Plugging these measurements into a published table with equations for different shark species then gives a highly accurate estimate of shark length.

How can sharks be repelled?

Perhaps this question should be rephrased as "Is there such a thing as an effective shark repellant?" The answer is yes and no. It depends on the shark and, more importantly, the circumstances. Curious sharks can be discouraged, and sharks that are leisurely feeding can sometimes be made to cease feeding, but actively feeding sharks (the ones we may be most concerned with) usually ignore any and all repellants.

Several natural chemicals that sharks appear to avoid or are repelled by have been tested. It has long been known that sharks avoid dead, rotting shark carcasses; the chemical or chemicals involved have been sought and tested, again with varying success. After World War II, the U.S. Navy provided packets of Shark Chaser, which consisted of cakes of what was thought to be the active ingredient in dead shark, copper acetate, combined with dark nigrosine dye. When dissolved in water by a downed pilot, the dye-chemical combination shrouded the diver at least visually and repelled any nearby sharks chemically. Tests showed that Shark Chaser worked about two-thirds of the time on a variety of sharks, but at other times, sharks were as likely to bite down on the cakes as avoid the area.

Other chemicals have been tried, again with varying success. Sea cucumbers secrete a substance called holothurin that results in rapid avoidance by some lemon sharks. A flatfish from the Red Sea, *Pardachirus marmoratus*, secretes a substance that sharks find repulsive (see "How do sharks find food?" in chapter 7). The active ingredient, a surfactant (like soap), and similar chemicals have been tested, although they have to be squirted directly at an approaching shark, an action that becomes tricky in an attack situation. Physical barriers have included a large, heavy-duty plastic bag with a floating ring on top developed by the U.S. military called the Shark Shield (less formally referred to as the whole-body condom). A swimmer, diver, or downed pilot climbs into the Shark Shield and theoretically decreases some of the signals given off that sharks find attractive. Other products include sonic guns and bubble curtains (variously repulsive or ignored).

A patent was granted in 1985 (U.S. Patent 4,494,245) for a shark-repelling wetsuit colored with boldly contrasting stripes or bars that mimic the warning coloration of sea snakes. The idea is that a shark seeing this would associate the color with that of the poisonous snake (and many other noxious animals) and therefore leave it alone. We contacted the patent holder,

a fish ecologist, and he informed us that he was not aware of any field tests of the concept. Given that many other shark repelling schemes have not worked, it at least seems worth trying.

Although most efforts at repelling sharks focus on reducing human injury risk, recent research has taken a different approach, one more concerned with shark welfare and conservation. Sharks are a large component of the "bycatch" associated with baited hook fisheries that target nonshark species, such as tunas and billfishes. The challenge, of course, is to discover something that deters sharks but doesn't reduce the attractiveness of the bait to the targeted fish. Some promising albeit not totally effective methods capitalize on the great sensitivity of elasmobranchs to electrical fields. Attaching magnets and electropositive metals to baits to produce strong electric fields has reduced catches of Atlantic Sharpnose Sharks, Smooth Dogfish, Blacktip Sharks, and Southern Stingrays but has had no effect on catches of Clearnose Skates or Spiny Dogfish, the latter a pest species in some fisheries (see "Are some sharks pests?" above). (As an aside, a neighbor of your first author had read that sending electrical impulses down his metal fishing line would increase his catch of salmon. Spiny Dogfish found it irresistible.) Catches of Sandbar Sharks were reduced in one study of bait magnets but not in another. Shark abundance also has an effect: solitary sharks may be repelled, but groups of sharks are apparently highly motivated to feed (see "Do sharks feel pain?" in chapter 8) and are not discouraged.

Personal safety devices using electrical shocks have been developed for divers, but a White Shark killed an Australian diver who was using one. The electrical current only served to hinder his mates from pulling him out of the water. Much obviously remains to be accomplished in the realm of shark repellants.

Do sharks have diseases and are they contagious?

Cruising the web, you are likely to find statements such as "Sharks are the only animals that almost never get sick: they are immune to almost every known disease." Sharks do appear to be immune to many diseases, including some that infect bony fishes. But sharks do have illnesses, including cartilage cancer and cancers such as melanomas and cutaneous fibrosarcomas (skin and connective tissue cancers).

Among the list of known shark ailments are a variety of bacterial diseases. The bacterial genus *Vibrio* includes species that cause cholera and others that cause gastrointestinal ailments often associated with undercooked seafood. A Sandbar Shark (*Carcharhinus plumbeus*) that died in a public aquarium was found to be infected with *Vibrio damsela*, focused in the kidneys. When injected experimentally under the skin of a Spiny Dogfish,

One variation on the patented shark-repelling wetsuit. The suit is decorated with bold, contrasting stripes to mimic the banding pattern of poisonous sea snakes. The varying angles of the stripes are to account for different swimming positions, because sea snakes have vertical black and white bands but a diver is not always lying flat in the water. Used with permission of N. Burkhead

V. damsela caused death in 18 hours. *V. damsela* can be transmitted to people, causing gastroenteritis and septicemia (wound infection). A captive Epaulette Shark (*Hemiscyllium ocellatum*) was found to be infected with *Mycobacterium avium*, a bacterium that causes coughing, fever, diarrhea, and weight loss in people. A dead Blacktip Reef Shark was found to be infected with haemorrhagic septicaemia, caused by the bacterium *Aeromonas salmonicida*. The bacterium caused inflamed lesions at the base of the fins and in the gills, liver, and intestine. *A. salmonicida* infects a variety of fish, especially salmon and trout, in which it is often fatal. Fortunately, it cannot be transmitted to people.

Sharks are immune to most viruses with the exception of a herpes virus found in wild and aquarium-kept Smooth Dogfish. This virus causes skin lesions that grow with time. It is thought that immunity to most other viruses is aided by the chemical squalamine, which sharks produce in their liver and other organs. Lab tests have shown that squalamine inhibits the replication of at least six different viruses, including those responsible for dengue fever, hepatitis B and D, yellow fever, and a form of encephalitis. Fortunately, squalamine can be produced synthetically and does not have to be obtained from live sharks (see "Are any medicines made from sharks?" in chapter 8).

Is it safe to eat sharks?

Generally yes (although not for the sharks). It all depends on the species and especially the place where it was caught. That latter fact is seldom if ever included on the label.

Sharks are top predators. This vaunted position in the food chain carries with it nutritional costs. Many chemical pollutants do not break down in animal tissues, meaning predators accumulate the dosage of their prey with each meal, adding to their "body burden" of pesticides, PCBs, PBDs, flame retardants, pharmaceuticals, heavy metals, and numerous other hazardous substances. Because sharks are predatory and live a long time, it is not surprising that their muscle, fat, and liver contain high levels of some pollutants.

Mercury is one such toxin. Sharks in some places have been found to contain mercury levels that exceed the safe level of one part per million (1 ppm) set by the U.S. Food and Drug Administration (FDA), and the more stringent (and precautionary) level of 0.3 ppm set by the U.S. Environmental Protection Agency (EPA). In Florida, fully one-third of the sharks sampled exceeded FDA standards. Shortfin Makos off New Jersey contained mercury levels of 1.8 ppm, almost double the FDA standard and six times higher than the EPA standard. Blue Sharks, Smooth Dogfish and Spiny Dogfish, three other species that find their way into seafood shops and restaurants, have also been found with levels of mercury that exceed the EPA standard. Some species, such as Smooth Dogfish and Spiny Dogfish, accumulate mercury as they age, rather than metabolizing it, meaning that larger individuals tend to be more contaminated. Fully 70% of shark muscle samples from dozens of species were at or higher than the maximum allowable limits for human consumption in the United States and Canada.

The liver is the major organ of decontamination in vertebrate animals, serving to remove toxins from an animal's blood stream. It is therefore not surprising to find elevated levels of pollutants in shark livers. Large, long-lived open-water and deepwater sharks appear to have the highest concentrations of liver contaminants; these include organochlorine pesticides (OCPs), which are synthetically manufactured chemicals like DDT, chlordane, dieldrin, and lindane. High levels of DDT and other OCPs have been found in the livers of Gulper Shark (*Centrophorus granulosus*), Blue Shark, and especially Greenland Shark. Both juvenile and adult White Sharks off South Africa have high DDT concentrations, not in their livers but in their muscle.

A study published in the *Marine Pollution Bulletin* found that Greenland

Sharks also contain high burdens of polychlorinated compounds (PCBs, dioxins). These compounds are known to cause "immunosuppression, embryonic abnormalities, reproductive dysfunction, carcinogenesis, and a severe loss of body mass" in vertebrates. Concentrations of these chemicals in the livers of Greenland Sharks were higher than in those of polar bears. Other sharks found to contain high levels of PCBs in liver and muscle include Common Thresher Shark, Bull Shark, Leafscale Gulper Shark, Tiger Shark, Blue Shark, Atlantic Sharpnose, and Smooth Hammerhead (*Sphyrna zygaena*).

Given these levels of contaminants in shark livers, it should be a matter of concern that shark liver oil is a source of vitamin A and a component in several pharmaceuticals. Liver oil from Nurse Sharks, Oceanic Whitetips, and Silky Sharks near Cuba showed PCB levels of over 14 parts per billion (ppb). A Japanese study also found PCBs and related polybrominated diphenyl ethers (PBDEs) in shark liver oil supplements. Canadian researchers also found PCBs and DDT in shark liver oil. Some health standards in the United States allow no detectable PCB concentrations of any level. PCB concentrations in Bull and Greenland Sleeper sharks at some locales are so high that a dead Bull or Sleeper shark found on a beach would have to disposed of as "hazardous waste." Just more reasons to not eat shark or ingest shark products.

We know that these pollutant concentrations can affect human health. What we don't understand is what they might be doing to the health of the sharks. The level of pollutants in Greenland Sleeper Sharks and Smooth Hammerhead Sharks greatly exceed the threshold amounts known to cause reproductive malfunction in aquatic mammals. We don't know what the critical levels of these toxins are in sharks, whether they are higher or lower than in humans.

In Iceland, a traditional food for social gatherings, called *Hákarl*, is made from Greenland sharks, a tradition that started long before the era of chemical analyses. The meat of these large sharks, which grow to over 6 m long and weigh 1,000 kg (21 ft and 2200 lb), is prepared by fermenting it (letting it rot) for several months. The dish is served in (fortunately small) bite-size chunks. Sleeper Shark muscle is very high in two nitrogen-rich chemicals, urea and trimethylamine oxide (TMAO). Eating a piece of *Hákarl* is like chewing on a sponge soaked in household ammonia (impression of first author, GH). The dish is made more palatable albeit still inedible by following each bite with a shot of a highly distilled vodka-like drink called *Brennivín*, or "black death," often accompanied by much singing.

Without fermentation (accomplished by burying the meat for several months), the nitrogen by-products in Greenland shark meat can be not

just unpalatable but also toxic. Reports in the literature indicate that sled dogs fed the raw meat stumble around, acting as if they were drunk. It's not known if the dogs were also given *Brennivín*.

What should I do if I'm attacked or injured by a shark?

Here's what the International Shark Attack File (www.flmnh.ufl.edu/fish/sharks/isaf/2012summary.html) recommends if someone is attacked by a shark:

> We advise a proactive response. Hitting a shark on the nose, ideally with an inanimate object, usually results in the shark temporarily curtailing its attack. One should try to get out of the water at this time. If this is not possible, repeat[ed] bangs to the snout may offer temporary restraint, but the result is likely to become increasingly less effective. If a shark actually bites, we suggest clawing at its eyes and gills, two sensitive areas. One should not act passively if under attack—sharks respect size and power.

Dealing with injury comes next. Minor bites and stab wounds from spines should be treated like any deep cuts, by stopping bleeding, cleansing the wound, using antibiotics to prevent infection, and so on. Some highly venomous stings, such as those caused by stingrays, can be excruciatingly painful. For stingray stings, first aid manuals suggest applying hot compresses (but not so hot as to burn) to the wound to destroy the proteins in the venom. Professional medical help may be needed because some people are particularly sensitive and can go into anaphylactic shock and stop breathing. The major danger from shark bites is blood loss and shock; even minor bites are likely to require prompt, professional medical attention.

Once you have received medical help, consider filing a report with the International Shark Attack File (www.flmnh.ufl.edu/fish/sharks/isaf/questionintro.htm).

Chapter 10

Human Problems (from a shark's viewpoint)

Are any sharks endangered?

The recognized worldwide authority on extinction risk in plants and animals is the International Union for the Conservation of Nature (IUCN), a United Nations–sanctioned clearinghouse for such information. The IUCN publishes a Red List of endangered plants and animals that it updates regularly. In 2012, the IUCN had assessed the status of 1,090 chondrichthyan species; population information was available for only half of these. When population trends indicate a species is threatened with extinction, the IUCN assigns ranks of "critically endangered," "endangered," or "vulnerable." The 2012 assessments indicated that 3% of sharks, skates, and rays were critically endangered, 4% were endangered, and 11% were vulnerable. This means that 18%, or about one-fifth of *assessed* species, are theoretically threatened with global extinction. Whether we can assume that unassessed species are equally at risk is an unknown.

That is the global view. At more local, regional levels, the picture can be even bleaker. Diver surveys at numerous Caribbean locales found few sharks other than Nurse Sharks. This conclusion stands in stark contrast to historical accounts of shark abundance in which researchers stated that a variety of sharks could be "expected anywhere at any time" in the west-Indian Caribbean. Today, except for Nurse Sharks, sharks are expected "anytime almost nowhere." In the Mediterranean Sea, comparisons going back to the early nineteenth century indicate significant declines in large sharks, especially hammerheads, Blues, makos, Porbeagles, and thresher sharks.

Human activities that have resulted in habitat destruction and pollution have undoubtedly affected shark populations. But overexploitation appears

to be the chief cause of decline in all these examples because it interacts with aspects of the basic biology of shark species that make them vulnerable to exploitation.

Because many sharks mature slowly and produce relatively few young at long intervals, they are easily overfished and their populations are slow to recover, if at all. Add to this the tendency of individuals to cross entire oceans, during which time they are exposed to fishing in multiple countries and on the high seas. Protective measures in one country do little to protect sharks when they travel to another region where protection is nonexistent. And the open ocean is largely unregulated and even harder to patrol.

Given these biological traits and circumstances, it is not surprising that commercial shark fishing has a long history of collapse. A typical shark fishery experiences initially high catches of a previously unexploited stock, followed by reduced landings despite increased effort, and eventual decline in catches to the point that it is no longer economically profitable to keep fishing. After most fishing ceases, populations are slow to recover, if they recover at all. Examples of sharks that have been overfished include Common Thresher Sharks and Pacific Angel Sharks in California; Angelsharks (*Squatina squatina*) in the Irish Sea; School Sharks and 12 species of deepwater squaliform dogsharks including Portuguese Dogfish off Australia; Spiny Dogfish in the North Atlantic; Basking Sharks off Ireland and British Columbia; Bull Sharks and Largetooth Sawfish (*Pristis microdon*) in Lake Nicaragua; Soupfin Sharks (*Galeorhinus zyopterus*) off the U.S. Pacific coast; Brazilian Guitarfish (*Rhinobatos horkelli*) in the southwest Atlantic; and Porbeagle Sharks off Newfoundland. Even chimaeras are subject to overfishing because, like sharks, they have a slow reproductive cycle. A fishery in New Zealand for Australian Elephant Fish collapsed in the 1980s from overfishing and has taken two decades to show signs of recovery.

The Porbeagle fishery of the western North Atlantic provides a classic example, collapsing after only six years. Exploitation began in 1961, with landings averaging 4,500 metric tons (MT; 9.9 million lb) per year prior to the collapse, when catches fell to around 350 MT (772,000 lb), or 8% of initial landings. Relaxed fishing led to partial recovery of this species, but only to about 10% to 20% of the original levels. Soupfin Shark landings off California peaked at 4,000 MT (8.8 million lb) in 1939, fell to 2,300 MT (5.1 million lb) in 1941, and five years later were down to 270 MT (595,000 lb), or 7% of the peak, despite intensive effort. Thirty years afterwards, populations were still below the pre-exploitation levels of the 1930s.

The history of the Basking Shark "fishery" off British Columbia is particularly sad because it involved the deliberate extirpation of a harmless giant species of no commercial value, eliminated solely because it was considered a nuisance to fishing interests. As slow-moving, plankton-feeding

A campaign button to promote conservation of Spiny Dogfish, produced by the Ocean Conservancy. An international effort was mounted to reduce allowable catch rates to preserve Spiny Dogfish stocks, especially in the Atlantic Ocean. New catch limits were established in 2000 for the Maine to Florida fishery. Once catch quotas are reached, fishing is supposed to cease. Photo by Gene Helfman

surface dwellers, Basking Sharks frequently became entangled in nets set or towed to catch other species such as salmon. Removing a 10-m-long (33-ft) animal from a net was inconvenient to fishermen, prompting a targeted campaign to rid the ocean of the inconvenience. Basking Sharks caught in nets were butchered and discarded, harpooned and discarded, shot for sport, and rammed and discarded by government boats specially equipped with knife-edged can openers on their bows. Records indicate that hundreds if not thousands of Basking Sharks were killed between 1920 and 1970. Since then, "only a handful of basking sharks have been sighted or caught . . . [representing] the first extirpation, in the true sense of the word, of a marine species" (Dulvy and Forrest, 656–57, in Carrier, Musick, and Heithaus 2010).

In most of these examples, the species became commercially rather than ecologically or evolutionary extinct. Commercial extinction occurs when a species falls to such low numbers that the effort that goes into catching the species makes it unprofitable to pursue, meaning that costs exceed profits. Commercial extinction precedes ecological extinction, which describes population levels so low that the species no longer functions in its natural role within an ecosystem. Commercial and ecological extinction are steps along the path to evolutionary extinction, or the death of the species. Springer and Gold (1989, p. 14) predicted this course for sharks when they wrote, "It seems certain that some species of sharks will be depleted and, perhaps, extinguished by man's activities."

In the ensuing two decades, their prediction has proven disturbingly correct. Shark and ray populations have crashed in many locales, and many species have been placed on endangered species lists, necessitating their protection from further exploitation. Some estimates from the North

Atlantic indicate that between 1985 and 2000, hammerheads were reduced 89%; thresher sharks, 80%; White Sharks, 79% (more than 6,000 captured); Oceanic Whitetips, 70%; Tiger Sharks, 65%; Blue Sharks, 60%; and several other species, 49% to 83%. Estimates from the Mediterranean put the declines of hammerheads, Blue Sharks, Porbeagles, Shortfin Makos, and thresher sharks at 96% to 99% of their former abundance. Reductions of this order at a global level would qualify these species for endangered or critically endangered status according to the IUCN criteria. The magnitudes of those estimates have been hotly debated in the scientific community, but biologists uniformly agree that the pattern of population reduction is globally widespread and of great concern.

Given these apparent historical patterns, fisheries for certain elasmobranch species have been increasingly regulated (some say too little too late), although many countries still lack any curbs on shark fishing. International protection has increased for Whale Sharks, Basking Sharks, and White Sharks, all three of which are protected from international trade by the Convention on International Trade in Endangered Species (CITES). In March 2013, six more species were given CITES protection: Oceanic Whitetip, Porbeagle, Scalloped Hammerhead, Great Hammerhead, Smooth Hammerhead, and both manta rays. The mantas, which are exceedingly slow breeders, have come under intense recent fishing pressure for their gill plates (see "Are any medicines made from sharks?" in chapter 8).

Whale Sharks are vulnerable because they do not occur in separate populations but instead make up one large global population. The entire world population of breeding-age Whale Sharks has been estimated at between only 27,000 and 180,000 individuals. Whale Sharks may be additionally vulnerable to exploitation because females tend to aggregate periodically in just two regions, the southern end of the Gulf of California in Mexico and the Galapagos Islands off the coast of Ecuador. This means that a significant number of females could be removed from the world population if fishing were allowed at those locales. Fortunately, the Galapagos Islands are one large marine refuge, and it is illegal to catch Whale Sharks in Mexico (although enforcement in both locations is hampered by a lack of funds).

Some species are exposed to extreme fishing mortality even though they aren't targeted by a fishery. Their size and anatomy cause them to be caught incidentally, a phenomenon known as bycatch. The giant Barndoor Skate of the northwest Atlantic and its relative the Blue Skate of the northeast Atlantic, ironically known also as the Common Skate, are caught in trawl nets dragged along the bottom that target shrimp, cod, or flatfish. The openings in the nets (mesh size) allow juveniles of most fishes to escape, but these rays are born at a size too big to slip through the nets. Both skates declined drastically in numbers and were thought to be threatened with

Visitors to the Georgia Aquarium in Atlanta view one of the two captive juvenile Whale Sharks there. Several countries have protected Whale Sharks, the world's largest fish, because of their great value in the dive tourism industry. Photo by Zac Wolf, Wikimedia Commons, http://commons.wikimedia.org/wiki/File:Male_whale_shark_at_Georgia_Aquarium.jpg

extinction. The northwest Atlantic Barndoor population is now thought to have recovered to 1960s population levels.

The most critically imperiled batoid group is the sawfishes (Pristidae), also victims of bycatch despite reduced targeted fishing. Sawfishes were once reasonably common throughout the waters around Mexico, Central America, western and eastern South America, most of Africa, Madagascar, the Arabian peninsula, India, Southeast Asia, the Mediterranean, the Philippines, and New Guinea. At one time, Lake Nicaragua contained a breeding population of Largetooth Sawfish, thought to have been the largest sawfish population in the world. Commercial exploitation began in 1970, with 60,000 to 100,000 sawfish caught over a five-year period. By the early 1990s, no more sawfish could be captured. Their capture finally became illegal in 2006.

Once found throughout the tropics in coastal areas with nearby rivers, sawfish ranges have shrunk drastically as the number of highly mobile adults has declined. Australia probably houses the largest populations, followed by the lower Amazon river basin and southern Florida. Although little directed fishing now occurs for sawfishes, they often become hopelessly entangled in nets of all types because of their tooth-studded snouts. Such bycatch remains the major threat to U.S. federally listed species such as the Largetooth Sawfish, which was once found in the Gulf of Mexico and ranged up the east coast to New York. Sawfish in the United States now occur only in extreme south Florida. According to some experts, sawfishes are arguably the most imperiled marine fishes in the world.

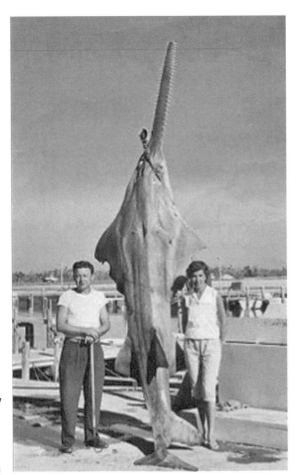

Largest officially measured Largetooth Sawfish captured in Florida waters. Sawfishes around the world are drastically overfished. Animals approaching this size (approximately 4.5 m, or 15 ft) are infrequently seen nowadays, and Largetooth Sawfish are now protected by federal law in the United States. Photo courtesy W. Bloyd

A Largetooth Sawfish swimming in the Ocean Voyager tank of the Georgia Aquarium. Photo by David Iliff, License: CC-BY-SA

When all fishes are surveyed for extinction risk, it becomes obvious that freshwater fishes have suffered more extinctions and population depletions than marine fishes. Human impacts, especially pollution and habitat destruction, have been more serious in freshwater because we use rivers and lakes as waste disposal grounds, withdraw water for consumption and irrigation, erect dams, dredge channels, and fish heavily. Not surprisingly, fresh- and brackish-water elasmobranchs as a group are disproportionately affected by human activities, putting many on endangered species lists. Of the 47 elasmobranch species that depend on freshwater or estuary habitats, 18 (38%) have been assigned to high-risk critically endangered, endangered, or vulnerable categories by the IUCN.

Will sharks be affected by global warming?

Global warming is just one aspect of global climate change. Climate change could influence many of the environmental factors that determine where sharks live. These factors include temperature extremes, oxygen availability, floods, droughts, major storms, and habitat loss. Some of these may have direct effects on sharks; others may affect the distribution and abundance of their prey, which will in turn impact sharks.

Global warming is likely to lead to alterations in wind direction and strength, which will, in turn, alter ocean currents. Ocean currents determine the timing and locales of spawning of ocean fishes, the dispersal of their larvae, and the distribution of adults. Altered currents could affect the distribution and production of open-ocean species that make up 70% of the world's fisheries, including the prey of open-ocean sharks. In the North Atlantic, where bottom temperatures increased only 1°C between 1977 and 2001, 24 of 36 fish species moved northward or deeper toward cooler waters, indicating that shifts in species distributions have already occurred. Because sharks follow their prey, distributional shifts in prey species could affect shark distributions.

Increased temperatures are also a threat because fishes often live close to the maximum temperatures they can tolerate, and because oxygen concentrations in water are reduced at higher temperatures at the same time that metabolic rates increase. Coldwater sharks may be squeezed towards higher latitudes to avoid warming ocean temperatures. Also, many pollutants are more toxic at higher temperatures. Corals are also very sensitive to slight increases in temperature, as was seen in 1998 when an increase of only 1°C in average temperatures killed 50% to 100% of the corals in some areas of the tropical Pacific. Many coral reef fishes are directly and indirectly dependent on live corals, and these fishes declined in many regions. Again, food availability for reef sharks could be an issue.

Are sharks affected by pollution?

As detailed in chapter 9 ("Is it safe to eat sharks?"), many pollutants wind up in the food chain and become concentrated at high trophic levels. Sharks are the top predators in many areas and eat prey animals already contaminated by a wide array of pollutants. We know very little about the direct effects of these pollutants on sharks, but given their effects on many other animals, it is likely they will have similar negative impacts on shark health, especially on large, long-lived species.

How are sharks affected by litter?

Because many sharks are scavengers, they tend to sample unlikely and unnatural items that are discarded at sea. Fortunately, sharks are able to regurgitate inedible items readily (see "What do sharks eat?" in chapter 7). One likely and unavoidable form of litter that has a known impact on sharks consists of discarded or lost fishing nets, especially monofilament gill nets. Sharks are often trapped in such nets when they blunder into them or try to feed on dead and dying fishes caught in the net. Sharks then twist and turn in an effort to escape these nets, thereby entangling themselves and losing the ability to move water over their gills, which leads to drowning. Other trash such as plastic straps encircle sharks and dig into their bodies. Monofilament fishing line wraps around the saws of sawfishes, cutting into and disfiguring the saw.

How do people catch sharks?

In modern times, most deliberately caught sharks are captured with baited hooks. Often, these lines—involving individual hooks or "long lines" with thousands of baited hooks—are actually set for billfishes, tunas, or other fishes; and the sharks are a bonus, or a nuisance. The sharks caught this way are processed for their fins and meat or simply discarded after their fins have been removed (see "What is finning?" below).

Some of the sharks that live near the bottom—the more abundant small sharks and most skates and rays—are caught in large trawl nets dragged along the bottom. The international fishery for Spiny Dogfish is one example, with some "net hauls" capturing thousands of animals.

In many places, sharks have been caught for food using a variety of traditional methods involving hooks made from bone or shell and various noise-making devices that attract sharks to a baited hook or within spearing distance. (See "What roles do sharks play in various cultures?" in chapter 11.)

Why do people hunt sharks?

At least 100 million sharks, skates, and rays are marketed annually; an equal number may die as bycatch in fisheries. The numbers are hard to nail down because landing statistics are collected carelessly and fisheries are regulated inadequately (see the next question). In addition, these numbers do not include subsistence and recreational fishing, the impacts of which have only recently gained the attention they deserve.

But sharks are and have been sought for a large variety of products and reasons. Shark meat is low in fat and provides a good source of protein for people living in many parts of the world, although contamination with pollutants such as mercury and various pesticides has become an issue in recent years (see "Is it safe to eat sharks?" in chapter 9). Shark liver oil is high in vitamin A and was a chief source of that vitamin before it could be synthesized in laboratories. Many pollutants are concentrated in shark livers, calling into question the advisability of using the oil as a food stuff. Fins are used to make shark fin soup (see "What is finning?" below). One popular item is stingray wings, often cut into circular pieces and passed off as scallops by less-than-scrupulous venders, though this practice attests to how tasty they are, since the ruse often succeeds. A large variety of useful drugs and other substances are derived from various shark products (see "Are any medicines made from sharks?" in chapter 8).

Sharks are also caught for pure sport. As large, active predators, many sharks put up a spectacular fight when caught with sport-fishing gear (others just pull slowly). Makos and threshers will jump when hooked. Shark fishing derbies were once common but have grown increasingly scarce as we learn how depleted shark populations have become in many places. Restrictions against shark fishing have grown stronger, to the point that sport fishers can only land and keep one or two sharks per day in most locales. Catch-and-release fishing has become increasingly common, but several shark species (hammerheads in particular) have low survival rates when released.

How are shark fisheries managed?

Poorly in most countries. Few countries have effective regulations affecting shark fishing. New Zealand instituted shark management in 1986, followed by Australia in 1988; the United States, Brazil, Canada, Costa Rica, the European Union, Mexico, Israel, the Maldives, and the Philippines have also established regulations, with many other countries considering legislation. Unfortunately, enforcement often lags far behind the passage of regulations around the world.

The United States lacked shark fishing regulations before 1993, when the National Marine Fisheries Service (NMFS) passed a Shark Fishery Management Plan (FMP) that limited the catch of 39 species in the Atlantic and the Gulf of Mexico. These regulations apply only to federal waters, which extend from 3 to 200 miles (5 to 320 km) offshore. However, most commercially caught sharks spend part or most of their lives closer to the shore than three miles, where control over fisheries is the responsibility of individual states, and states were slower in regulating shark fishing. The FMP did not regulate Pacific states, including Hawaii, which has the highest shark bycatch rate and where 61,000 sharks were finned in 1998 alone.

The 1993 FMP set a fishing limit of 2,436 MT (5.4 million lb), but that limit proved biologically unrealistic: shark populations continued to decline. In 1997, the quota was cut in half, to 1,285 MT (2.8 million lb); more species were added to the list; recreational fishing was reduced from four to two sharks per day per boat; and complete bans were instituted for Whale, Basking, White, Sand Tiger, and Bigeye Sand Tiger sharks. Because of a lack of data on the extent of the fishery and the biology of many species, the fishery was divided into three management groups instead of placing limits on individual species. The three groups were large coastal species (e.g., Tiger, Lemon, hammerheads, Blacktip, Bull), small coastal sharks (Sharpnose, Finetooth, Bonnethead, Angel), and pelagic sharks (makos, Blue, Porbeagle, threshers, Oceanic Whitetip). This "aggregation" of species into groups ignores biological differences. It's sort of like setting hunting restrictions on deer by counting deer, pigs, turkeys, ducks, and bear as a single group. Not surprisingly, achieving management goals was difficult. But if sharks were being managed partially out of ignorance, at least they were being managed.

The NMFS recognized these shortcomings and proposed numerous improvements in data collection and management goals, particularly among open-ocean and large coastal species, and placed restrictions on the capture of several other skate, sawfish, and shark species. Overall, the modified FMP has been recognized as a good example because of its relative thoroughness; and U.S. fishery scientists have been active advocates for improved monitoring, management, and cooperation in regulating shark fisheries around the globe.

Despite evidence that many shark species are in decline, and despite the efforts of dedicated biologists and others, management of shark fisheries at the international level remains ineffective or nonexistent. More so than most other fisheries, shark fisheries have to be managed simultaneously at local, national, and international levels, greatly complicating efforts. Sharks cross management boundaries and often enter waters where no one

has control and no restrictions apply. International governing bodies have, however, failed repeatedly to institute needed restrictions. It was only in 2004 that bans on the sale of Whale, White, and Basking sharks between countries were finally instituted.

Some regulatory action is likely to come from the U.N. Agreement on the Conservation and Management of Straddling Fish Stocks and Highly Migratory Fish Stocks, which was formalized in 1995 and is slowly being ratified by member nations. In 1999, the U.N. Food and Agriculture Organization (FAO) produced an International Plan of Action for the Conservation and Management of Sharks (www.fao.org/waicent/faoinfo/fishery), another positive step. Member states were encouraged to develop national plans of action that included regular assessments of shark stocks and data sharing among nations, particularly for migratory species. However, obstacles to improvement remain substantial. For example, India, the world's most active shark fishing nation, reports catches solely as "elasmobranchs," a group that includes 20 shark species and 45 skates and rays. But as we have seen, sharks differ in abundance, life history characteristics, habitat, feeding, behavior, and, consequently, vulnerability to exploitation. Managing elasmobranchs as a group makes little sense.

To date, action by U.N. members has been slow at best, although the situation appears to be improving with time. Another cause for hope is the large and dedicated constituency of people and organizations who care about sharks. This group includes the American Elasmobranch Society, the Shark Specialist Group of the IUCN, the Shark Trust in the United Kingdom, the Oceania Chondrichthyan Society, and the European Elasmobranch Association, as well as numerous conservation organizations. (See appendixes B and C for more information on shark conservation organizations and their work.) Our hope is that readers of this book will join in these efforts to protect and conserve these magnificent animals.

What is bycatch and can it be reduced?

Bycatch, sometimes referred to as by-kill, involves the incidental capture of sharks in fishing operations that were targeting other species. Because of the feeding habits of sharks, they are often caught on baited hooks, regardless of what the hooks were actually meant to catch. Sharks, skates, and rays become entangled in gill nets and are often taken in trawl nets designed to catch other species.

Because bycatch is often not reported, it is difficult to know with accuracy just how many sharks are caught incidentally in the world's fishing industry. Almost no data are available for recreational, subsistence, and artisanal fisheries, adding to the inaccuracies. About 100 million sharks,

skates, and rays are caught annually in targeted fisheries (including 30 to 70 million sharks for the fin trade), and an equal number may be caught and never reach a market. On a weight basis, about 760,000 MT (1.7 billion lb) of chondrichthyans are reported as caught annually, but that number probably doubles to 1.5 million MT (3.4 billion lb) when the unreported bycatch numbers are added. Few fishing operations go to any trouble to release sharks alive because sharks caught in other operations that aren't finned are viewed as nuisances, using up bait intended for more valuable fishes, tangling lines, and slowing the fishing operation. Fishermen take out their frustration on the sharks.

Reducing bycatch therefore benefits fishers as well as sharks. The trick, of course, is coming up with a fishing technique that catches the more desirable, targeted species while repelling sharks. One of the more promising methods being tested is the incorporation into baited hooks of magnets and electropositive metals, which sharks find undesirable (See "How can sharks be repelled?" in chapter 9).

What is finning?

In some places, fishers remove the fins from sharks and discard the carcass, now referred to as a "log." The finless shark, sometimes still alive, then sinks to the bottom to slowly starve to death or to be killed by predators and scavengers.

Shark meat has value, but shark fins are one of the most expensive forms of seafood around. Depending on species, shark steaks and fillets can cost a few dollars a pound or as much as $22/kg ($48/lb), although concern over high mercury content affects the marketability of some species. In contrast, the fins of sharks and large rays, when dried, processed, and sold in Asia, can cost as much as $1,650/kg ($3,630/lb). This makes shark fins one of the world's most valuable fishery products, worth almost twice as much as the most expensive Bluefin Tuna and approaching the value of fine caviar.

Dried fins are used in the preparation of shark fin soup, popular at wedding banquets and business dinners in Hong Kong. Ironically, the soup may contain little actual fin, this ingredient having been minimized to keep the price affordable (a bowl may sell for as much as $150). The cartilage supports in the fins, referred to as "noodles," are used primarily as a thickener, the flavor coming largely from chicken or other meat stock. The soup may even be filtered before serving and the solid fin parts discarded.

Hong Kong imported 7,800 MT (17.2 million lb) of shark fins worth US$250 million in 1996. Since fins account for only 4% of the body weight of a shark, approximately 200,000 tons, or 440 million pounds, of sharks were killed in the process. Hong Kong is said to account for half of the

Shark fins bound for Chinese markets are dried and processed in Thoothoor, India. Photo courtesy of Jeff Rotman, www.jeffrotman.com

global shark fin trade, so you can do the numbers. And that reflects only the *reported* catch. We have no idea how large the black market may be.

Shark finning is now banned in the United States, following enactment of the Shark Conservation Act, signed by President Obama on January 4, 2011. The act protects most shark species from being harvested for their fins by prohibiting the cutting off of fins at sea, and it prohibits the landing of shark carcasses without the fins being "naturally attached." The law allows shark fin suppliers to sell the product if the fin is cut off after the shark has been landed dockside, so selling fins is still legal.

International outcries over the brutality of finning have led to a groundswell of action to discourage and outlaw the activity globally. Several organizations have spearheaded campaigns to educate the public about the process, with messages put out by just about every shark conservation organization listed in appendixes B and C. As a result of these efforts, the number of countries that have outlawed finning has increased substantially in recent years, and sales of shark fins and shark fin soup have declined steadily. Hong Kong government statistics show that although imports of fins continued to average around 10,000 MT (22 million lb) between 2006 and 2011, a sharp drop to 3,350 MT (7.4 million lb) occurred in 2012. The Shark Fin Trade Merchants Association reported a similar 60% reduction in sales of dried fins. Grass roots efforts can make a difference!

Is there such a thing as shark leather?

Shark skin has been tanned to make leather for centuries because of its toughness and flexibility when properly cured. ("Sharkskin" business suits are not made from sharks; their fabric is called sharkskin because of the shine of the cloth.) An active fishery for shark hides existed in Florida in

STOP SHARK FINNING

KEEP SHARKS IN THE OCEAN AND OUT OF THE SOUP

Part of the anti–shark finning campaign promoted by www.stopsharkfinning.net. Several hard-hitting videos on the subject can be found by searching for "shark finning" on Google videos or YouTube. Image from www.stopsharkfinning.net

the 1940s, when hides were divided into "nurse shark hides" and "other sharks." The more desirable Nurse Shark hides were worth about 25% more than the others. For ages, dried shark skin, called "shagreen," was used as a high-quality sandpaper, at least until the glues and papers of modern sandpaper were invented. Aleut Indians (Eskimos) use dried shark skin for making boots. The handles of Japanese samurai swords were sometimes wrapped in shark leather to make them less slippery; the denticles in the shark skin provided a better grip if the handle became blood-covered.

Shark and ray leather wallets, belts, coin purses, and other items are available through many catalogue and Internet sites. Stingray leather wallets often include the region of the barb highlighted in a different color, adding to the novelty value of the article.

Is there such a thing as shark jewelry?

Just about every curio shop in the world, but especially those located near a beach, seems to sell shark teeth jewelry. Often it's just a tooth with a string or leather thong attached, although sometimes the base of the tooth is wrapped in wire. If it's a recent and not a fossilized tooth, it may have a hole drilled in it (fossilized teeth are actually rocks and are much harder to drill). Shark tooth jewelry is also sold on numerous Internet sites. Historically, fossil shark teeth were worn as amulets or carried in pouches to ward off evil and to detoxify poisons (see "Did early philosophers and naturalists mention sharks in their writings?" in chapter 11).

Occasionally stingray barbs are made into jewelry. We've seen large barbs fashioned into letter openers, set in a leather handle, or even set in a handle fashioned from hawksbill turtle shell. Sawfish saws with all the teeth attached are still a sought-after object in the curio trade, with large, intact saws bringing over $1,200. Such trade is now illegal because of protection in Appendix I of the Convention on International Trade of Endangered Species (CITES), although thousands are still sold for use in Chinese traditional medicine. eBay supposedly discontinued listing sawfish saws as of 2006, but they still appear there from time to time. In any case, tooth or barb, the object is still sharp and should be treated as such.

Sharks: The Animal Answer Guide

A stingray leather coin purse. The brighter, oblong area in the center is said to come from the base of the ray's barb. Stingray leather is exceedingly durable. Such coin purses and wallets sell for $50 to $80. Photo by Gene Helfman

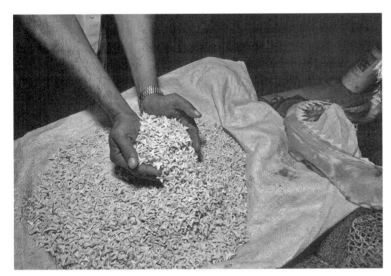

Shark teeth are sold individually to tourists in curio shops around the world. A single tooth in a small plastic bag, sometimes with the species identified but often not, usually sells for a couple of dollars. Here, a shark fisherman in Thoothoor, India, loads shark teeth into a burlap bag for later sale. Needless to say, the fisherman receives much less than a dollar a tooth. Photo courtesy of Jeff Rotman, www .jeffrotman.com

What other parts of sharks are used by humans?

Shark skin (as leather), shark teeth (jewelry, weapons, and tattooing tools), sawfish saws (fashioned into battle weapons), and stingray barbs (knives) are just some of the parts of sharks that have been used over the centuries. There have also been some more unusual uses. Alaskan Inuits have used the lower jaws (dental bands) of somniosid sleeper sharks as knives for cutting hair, and various cultures have used shark oil barometers. In Bermuda, shark livers were boiled to render the oil, which was then placed in a clear bottle. When an approaching storm caused the barometric pressure to drop, the shark liver oil would change from a clear golden color to milky white. Cockfighting is popular in some Asian and Latin American

cultures, and the human participants attach sawfish rostral teeth to the legs of the roosters as artificial spurs.

Are boats dangerous for sharks?

Most sharks are sufficiently mobile to get out of the way of an approaching boat. However, some of the bigger plankton-eating elasmobranchs do not have a natural fear of large, fast-moving objects and are struck and injured by boats. Manta rays and Whale Sharks have been struck by boats, and Basking Sharks were deliberately rammed and cut up during a twentieth-century effort to rid the northeastern Pacific of these planktivores. Some accounts report speedboats using Basking Sharks as ramps to be jumped (see "Are any sharks endangered?" above). Stingrays that are attracted to ecotourism feeding sites in artificially high numbers, such as at Stingray City in the Cayman Islands, are frequently injured by boats as the rays jockey for position to be fed (see "Should people feed sharks?" in chapter 8).

What can an ordinary citizen do to help sharks?

A lot! Some actions are quite simple:

- Refuse to eat sharks, and make your refusal known at restaurants and seafood counters. Encourage your friends and family to also refuse to eat sharks. Some shark populations can sustain well-managed fisheries, but most can't, and seafood sellers seldom know where their sharks come from and whether they were caught sustainably. There are many other tasty kinds of fish on the menu. To know which ones to avoid or consume, consult the readily available Seafood Watch guides such as the ones produced by the Monterey Bay Aquarium (www.seafoodwatch .org).
- Support anti-finning campaigns. Write letters to politicians and government agencies supporting all bans on shark finning.
- Discourage shark fishing "kill" derbies by writing letters to local papers pointing out that many of the sharks caught in these "contests" are declining in numbers. Such derbies are commercial enterprises organized to promote tourism and sport fishing. Businesses hate the negative publicity created by a well-written letter to the editor of a local newspaper.
- Most importantly, stay informed. The avalanche of information on sharks and shark conservation grows annually. It may appear hard to keep up, but a few sites troll the scientific literature and media sources and summarize current information. If you want access to the best sci-

ence on shark conservation, check out the websites of the IUCN Shark Specialist Group (www.iucnssg.org) and the Florida Museum of Natural History (www.flmnh.ufl.edu/fish), the latter maintained by the second author of this book (GHB) and his dedicated staff. Other useful sites include the Global Shark Conservation Campaign of the Pew Charitable Trusts (www.pewenvironment.org/campaigns/global-shark -conservation/id/8589941059) and Shark Year Magazine (http://shark year.com).

Chapter 11

Sharks in Stories, Media, and Literature

What roles do sharks play in religion and mythology?

The sea is a dangerous and mysterious place and an understandable source of myths and legends. Sharks undoubtedly played a part in the stories of "sea monsters" who gobbled up sailors. Without photographic equipment to document a brief, tragic event, descriptions of the perpetrators were likely to be inaccurate if not embellished, fueled by the imaginations of sailors already strained by the perils of long ocean voyages. Thus were born various sea serpents. Some sightings are now thought to have been decomposing Basking Sharks. Baskers grow to 12.5 m (40 ft) in length and leave behind a cartilage skeleton with a relatively small head, a long neck, and a long tail. Beached, decomposing Basking Sharks still show up in modern news accounts as "mysterious sea creatures," or even reptilian plesiosaurs, until a knowledgeable ichthyologist dispels the mystery.

Seafaring nations, from small tropical islands to continental locales, all respected the sea and its denizens and developed rituals to placate ocean deities or ensure harvests and safety. One source of legend, myth, and worship in many countries are the very real pristid sawfishes. Sawfishes occur, or occurred, globally in tropical marine nearshore areas. They often enter fresh water, which means people did not have to go to sea to encounter them. At lengths of over 7 m (23 ft), weighing in excess of 2,250 kg (5,000 lb), and equipped with a swordlike snout lined on either side with inch-long (2-cm) spiky teeth that they use for slashing, they demand respect.

Examples of sawfishes in story and culture come from across the globe. The Aztecs believed that a monstrous sawfish named Cipactli was responsible for forming the earth and kept crops growing. Cipactli had to be fed

Aluminum-bronze 5-, 10-, and 25-franc coins in circulation in the West African (CFA) countries of Benin, Burkina Faso, Cote d'Ivoire, Guinea-Bissau, Mali, Niger, Senegal, and Togo. Each coin bears a highly stylized pristid sawfish on its face, although the uneven rostral "teeth," slender body, and large tail fin are also reminiscent of the pristiophorid sawsharks. Exchange rates run at about 500 CFA francs to the dollar. Wikimedia Commons, http://en.wikipedia.org/wiki/File:BCEAOFranc.png

human blood regularly to maintain her strength. These sacrifices included crushing the victim's neck with a sawfish snout, symbolizing Cipactli's feeding behavior.

In Asia, traditional healers chased away demons and cured disease by wielding a sawfish saw. In northern Borneo, locals traditionally nailed saws over doors to repel ghosts, or hung wrapped saws over cradles to soothe infants. In Papua New Guinea, sawfish are said to unleash storms to punish those who break fishing taboos.

Several African coastal societies viewed sawfishes as one of the most dangerous beasts around, attributing to them supernatural powers to repel evil, cure sickness, and overcome misfortune. The Lebou of Senegal protect their homes, families, and cattle by mounting saws on the walls of their homes. Sawfishes find their way into proverbs of the Duala of Cameroon. The Jola of Gambia and Senegal view sawfish as ancestors possessing a magical weapon. To the Akan of Ghana, the sawfish is a symbol of the authority of the king and a guarantor of the kingdom's prosperity. A continued high regard for sawfish in African societies is reflected in the coinage of the West African states.

The Kuna people of the San Blas Islands of Panama also revere sawfish. Legends tell of sawfish patrolling the coast, protecting fishermen from sharks and other dangerous animals and even rescuing drowning fishermen.

Myths and legends involving other sharklike fishes occur in many cultures. Many myths tell of giant fishes that are responsible for the creation of nations and the heavens, that help hold up the world, or that are responsible for earthquakes, tidal waves, erosion, and strong winds. A Maori legend from New Zealand tells how Ikaroa, a long fish (perhaps a shark), gave birth to the stars in the Milky Way. A Japanese giant fish named Isonade is a shark-like monster that causes strong winds, capsizes boats, or uses hooks on its tail to catch unwary fishermen. A Kuna story describes the activities

of a spirit named Manta Ray Man, a violent drunk. Manta Ray Man started fights by flailing and flapping his arms the way that manta rays flap their "wings." Respect for manta rays and their power has shown up across cultures and centuries. For example, pottery unearthed in Peru and dated to over 1,800 years old contains obvious manta ray images.

Throughout the Pacific, island cultures were dependent on the sea, with much local art, wisdom, and lore focused on sea creatures such as sharks and their relationships with humans. Sharks still show up in contemporary art forms in many nations. In the Palau Islands of Micronesia (Republic of Belau), legends say that when a shark swims upside down and bites your canoe, your wife is cheating on you. Many such stories were carved into the walls of the men's meeting houses or onto storyboards, now sold to tourists.

Native Hawaiians have always respected sharks, known collectively as *mano*. Many aspects of Hawaiian history and culture reflect this reverence. Hawaiians recognized a number of animal deities or gods who were thought to embody their ancestors. Among the family deities (*amakua*), sharks are considered the strongest. Families that had shark amakuas would neither eat nor catch sharks. Fishing for the largest and most dangerous *niuhi* sharks, namely Tiger and White, was an activity reserved for chiefs. Eating the eyes of niuhi brought mystical powers and strength. The mother of King Kamehameha I ate niuhi eyes to give her unborn son greater leadership qualities and bravery.

An oft-repeated Hawaiian legend relates how the king of sharks fell in love with a human woman, to whom was born a shark-human son named Nanave. Nanave swam daily in a pool connected to the ocean. He often asked passing fishermen where they were going to fish and would then vanish into the pool to warn his fish kin of their peril. Upon discovering that Nanave was part shark, the villagers attempted to kill him, but he escaped into the pool. From that day, Hawaiian fishermen have not revealed where they are planning to fish. As with many legends from across the globe, the story of the king of sharks may explain a natural phenomenon. Nanave's tale may reveal that an absence of fish on the reef could be due to a large shark patrolling the area, frightening off other fish.

Cultures outside the tropics also incorporate large sharks into their mythologies. The Inuit of Greenland hunted *Skalugsuak* (Greenland Sharks), one of the world's largest predatory sharks (up to 6.4 m, or 21 ft, long). Greenland Shark meat contains high concentrations of urea and must be processed correctly to be edible (and even then, it's an acquired taste). An Inuit legend recounts the origin of the Greenland Shark. An old woman washed her hair in urine, dried it with a cloth, and when the cloth blew into the ocean, it became Skalugsuak. Another Greenland legend tells of Sedna,

a girl whose father cut off her fingers while drowning her. Each finger became a sea creature, one of which was Skalugsuak.

New World temperate societies also had legends about sharks and their relatives. The Snununeymuxw people of British Columbia (www .snuneymnuxw.ca) tell of Stleluqum, a huge stingray-like creature that lived in the narrow, turbulent channel between Newcastle Island and Nanaimo, British Columbia. Stleluqum's activities caused the water to boil, and the Snununeymuxw referred to the place as *Lhapqwum*, or "boiling place." One day, a young man tried to swim across the channel to visit a young woman with whom he had fallen in love. Stleluqum swallowed him, just as the young man had been warned. His lover screamed, and her relatives came running. When she explained what had happened, the village elders advised the young people to gather clams and their spears and paddle out to the monster. When they tossed the clams (a favorite food of rays everywhere) into the water, the monster came up, and they speared and killed it. When they cut it open, the young man jumped out unharmed. (One moral of the story is that young people should respect their elders and learn from them. The authors of this book are much older than you.)

What roles do sharks play in Western religions?

According to Judeo-Christian scripture, sharks (and other aquatic organisms) were created on the fifth day. Genesis 1:20–21 of the Old Testament (New American Standard Bible, 1995 update) reads: "Then God said, 'Let the waters teem with swarms of living creatures. . . .' [and] God created the great sea monsters and every living creature that moves, with which the waters swarmed after their kind."

Sharks as such do not appear to have been mentioned in either the Old or New Testaments, which isn't surprising, given how little Westerners knew about large sharks before the sixteenth century (see "How did sharks get their name?" in chapter 1). However, many interpretations of the story of Jonah involve a large shark rather than a whale, a point argued by biblical scholars as well as ichthyologists for centuries. The original text (Jonah 1:17), literally translated, reads, "And the lord provided a great fish for Jonah, and it swallowed him, and Jonah sat in the belly of the fish for three days and nights." It wasn't until 1534 that the phrase "great fish" was translated to mean a whale, which became the authorized version in 1611. No small amount of the confusion arises from the fact that whales were thought to be fish at that time. This was corrected in 1758 when Swedish biologist Carl Linnaeus, in the tenth edition of *Systema naturae*, recognized whales to be mammals. Given the close proximity of biblical authors to the Mediterranean Sea, Jonah's attacker was probably a White Shark. This

might make Jonah 1:17 the first record of a shark attack in Western history and therefore case number 1 in the International Shark Attack File.

Sawfish, referred to as *Serra*, appear as early as the tenth century in medieval bestiaries. These natural history guides, dating back to the second century, were generally lavishly illustrated and contained often-fanciful descriptions of animals. In later centuries, accounts were linked to lessons, parables, and passages of the Christian Bible. *Serra* has the same root as *serrated*, meaning "sawlike." The serra of the bestiaries are said to have raced sailing ships unsuccessfully, although some versions have them attacking boats. The illustrations accompanying the accounts bear no resemblance to actual sawfish, however.

Did early philosophers and naturalists mention sharks in their writings?

Pliny the Elder (23 CE–79 CE) was the first Roman philosopher and natural historian to describe fossil shark teeth. He surmised that they fell from the sky during lunar eclipses. Later natural historians who found large fossil teeth speculated that they were the petrified tongues of dragons and snakes. These *glossopetrae* (literally, "tongue stones") were considered a remedy for various poisons and toxins, including snake bites. Royalty wore glossopetrae pendants or kept them in pockets and purses as good-luck charms. Their actual nature wasn't understood until 1667, when Danish anatomist Nicolaus Steno dissected a White Shark's head and illustrated the teeth; the link to glossopetrae became obvious. Necklaces and pendants made with fossil shark teeth remain a popular form of jewelry, regardless of purpose (see "Is there such a thing as shark jewelry?" in chapter 10).

Aristotle (384–322 BCE) is considered by many to be the father of zoology. He was the first (among Western philosophers) to make careful observations and dissections, at the time a revolutionary act. Among his discoveries were that some sharks produce eggs that hatch within the mother, and that some sharks have a placenta-like structure and an umbilicus (see "Do sharks care for their young?" in chapter 6). He also published detailed descriptions of the anatomy of electric torpedo rays. Aristotle's dissections of torpedo rays were relevant because these marine electric fishes, which produce strong electric shocks equivalent to about 1 kW of power, were used as shock therapy to cure a variety of ailments, including epilepsy.

What roles do sharks play in various cultures?

It is no surprise that sharks played and still play an important part in many cultures. The power of large sharks and their potential to injure and

Ulisse Aldrovandi's apparent rendition of a sawfish, circa 1613. According to an early-twentieth-century historian, "It would be hard to find any creature treated by artists in more diverse fashion, and it is frankly evident that none of them knew what it was like, if indeed it was to be seen in the flesh at all." A gallery of other, equally imaginative early portrayals of sharks and sharklike animals can be seen at www.strangescience.net/stsea2 .htm. *De piscibus libri V et De cetis lib unus. Bononiae*, [Bologna, Italy]: Apud Bellagambam

kill inspired stories, legends, and myths. Sharks are also a source of protein that can (or previously could) be caught year round, although subduing large sharks certainly required courage and skill. Specially constructed large shark hooks were common. Devices for luring sharks took advantage of the way that repetitive, low-frequency sounds attract sharks (see "How well can sharks hear?" in chapter 2). In the Pacific, fishers made rattles from multiple coconut husks strung together, lowered them from a canoe, and shook them, perhaps mimicking the sound of an injured fish. Sharks and stingray parts have also been used for leather, tools, and weapons (see "Is there such a thing as shark leather?" in chapter 10). Stingray spines in particular have made effective awls for poking holes in leather, as well as serving as knives and daggers.

Shark teeth have been widely used as tools to carve wood and prepare food; as weapons attached to clubs, daggers, and swords; as spear or arrow points; for skin-piercing tools to create tattoos; and as ceremonial objects. Easter Island, home of the famous *moai* stone statues, is also famous for wooden *rongorongo* tablets covered in hieroglyphics (picture writing) that were apparently carved with shark's teeth. Native Hawaiians produced formidable war clubs edged with shark teeth, especially from Tiger Sharks. These *leiomano* included a variation on brass knuckles but edged with shark teeth and clubs edged with shark teeth. Shark teeth and stingray spines augmented the destructiveness of weapons in other Polynesian cultures such as on Kiribati (central tropical Pacific) and Nauru (Micronesia), where stingray skin was also used as a type of armor worn by men in battle. (Women also fought in such contests but apparently weren't allowed to wear armor.) Shark teeth also found their way into Japanese culture, where traditional

Sharks in Stories, Media, and Literature

Useful shark's teeth. (*Left*) Actual wooden weapons studded with shark's teeth, from the Gilbert Islands in the Republic of Kiribati. (*Right*) A modified *Carcharocles megalodon* fossil tooth offered for sale on eBay. The description claimed that it was the "Oldest Tool Made by Humans in North America—Megalodon 20,000 Year Old Spear Tip." Asking price: $995,000. *Left*, http://en.wikipedia.org/wiki/File:Shark-tooth_weapons.jpg; *right*, eBay item 140511853253

Japanese female entertainers (geishas) used crushed Tiger Shark teeth to produce the white pigment with which they painted their faces.

Shark teeth were a valued trade item among Native Americans. Shark tooth artifacts have been found in the 700- to 1,000-year-old Cahokia mounds across the Mississippi River from what is now St. Louis, Missouri—more than 1,000 km (620 miles) from the ocean. Such objects have also been found in Ohio burial mounds dating back 1,700 to 2,100 years ago, as well as in sites as old as 4,000 years in northeastern U.S. coastal states and the maritime provinces of Canada. Native Americans used both fossilized and recent (nonfossilized) shark teeth and vertebrae as tools and especially as ornamental items. The teeth and vertebrae often had holes drilled in them, suggesting they were strung on necklaces or hafted (attached) on wooden clubs, spears, or paddles. The burial sites are typical of those of clan leaders, and the inclusion of shark parts attests to their value as traded objects. Anthropologists call such objects "prestige goods." Items from distant locales made from exotic materials such as marine shells, copper, nonlocal flint, or shark parts increased the status of the owner. And, after all, life is really about one's image.

What roles do sharks play in popular culture?

Sharks are everywhere nowadays. They turn up in movies, TV shows, books, cartoons, music, video games, license plates, weather vanes and windsocks, boat buoys, postage stamps, pinball machines, T-shirts (we have dozens, some of the best found at www.trollart.com), wallets and purses, towels, stuffed animals, Halloween masks, refrigerator magnets, doormats, house signs, and beer brands. All attest to their popularity with the public.

A White Shark's lower-jaw tooth found at the Cahokia archeological dig near St. Louis, Missouri, dating back perhaps 1,000 years. The hole drilled in the tooth suggests it had been strung on a necklace or attached to a wooden handle as a weapon or tool. Given the tooth's size—2 cm (0.8 in) wide and almost 3 cm (1.2 in) tall—it probably came from a fairly large animal. Photo © Ken Farnsworth, curated at the Gilcrease Museum; photo provided courtesy of Laura Kosuch, used with permission of copyright holder

In many cultures, it is believed that one can attain the strength, courage, or other desirable characteristics of an animal by carrying or using an object directly tied to it, like wearing a shark tooth necklace. Considering this, we guess that Great White ale makes sense. More puzzling is the Dogfish Head variety of beers. Photo by Gene Helfman

Movies. Hollywood and its foreign wannabes know sharks sell and consequently churn out big and especially low-budget shark attack movies. These films are often as not set in undersea labs or at spring-break or other vacation beach locations, involving ginormous psychomutant/genetically modified sharks, scantily clad coeds, and marine biologists (some female and also scantily clad) as both villains (mad scientist types) and heroes. *Jaws*

(1975) started it all, although none of the movies that followed, including the *Jaws* sequels, had the same impact at the time they were released. Few shark movies make it past the small screen and the Syfy channel; several appear on the "50 Worst Movies Ever Made" lists.

Just about every habitat imaginable has been terrorized by blood-thirsty selachians, as reflected in the titles of three movies released in 2011: *Snow Shark—Ancient Snow Beast*, *Sand Sharks*, and *Swamp Shark*. (We don't know of any sharks-from-another-planet movies, but we wouldn't be surprised.) Sometimes the plot varies slightly and only the location changes, as in *Shark in Venice* (2008), *Malibu Shark Attack* (2009; at least this one featured Goblin Sharks, instead of the usual White/Bull/Tiger/Mako mutants), and *Jersey Shore Shark Attack* (2012).

Certain other themes are rehashed repeatedly. Some give us an inkling of what the villain will look like, as in *Tintorera Tiger Shark* (1977), *Great White* (1981), *Megalodon: Fear the Water* (2002), *Hammerhead: Shark Frenzy* (2005), *Mega Shark vs Giant Octopus* (2009), followed by *Mega Shark vs Crocosaurus* (2010), *Sharktopus* (2010), and *Two-Headed Shark Attack* (2012). Others just hint at anatomical excess; examples are *Devilfish/Monster Shark* (1984), *Dinoshark* (2011), *Super Shark* (2011), and *Jurassic Shark* (2012). A few assure us that no red dye nor screaming victims were spared, as in *Deep Blood* (1989), *Cruel Jaws* (1995), *Redwater* (2003), *Raging Sharks* (2005), and *Blood Surf* (2001—spread blood and guts on the water, then surf through the feeding frenzy that results). *Spring Break Shark Attack* (2005) didn't fool around, beginning with its title. A few could involve any of Hollywood's backlot closet of monsters, but don't worry, as in *12 Days of Terror* (2004) and *The Reef* (2010).

Spoofing the sharksploitation genre is *Ghost Shark 2: Urban Jaws* ("And for God's sake, stay out of the water!"). This much-delayed and -anticipated movie has a 2013 release date; it even spoofs itself because there never was a *Ghost Shark 1*. Somewhere between spoof and serious (?) is *Sharknado*, the Syfy channel's 2013 twist on the sharkastrophe theme (pun intended). Bloodthirsty sharks are picked up in a waterspout/tornado and deposited on the unsuspecting summer crowds of Los Angeles, including at an amusement park where the patrons get more of a thrill ride than they bargained for. The movie includes the memorable rip-off line, "We're gonna need a bigger chopper." And after *Sharknado*, is it not reasonable to expect *Sharkano*, *Sharquake*, and *Sharricane*?

Among big-screen, big-budget Hollywood productions that contain if not feature sharks is *The Perfect Storm* (2000), with its spectacular swordfishing scenes that included cameo performances by Blue and mako sharks (and also some bad weather). *The Life Aquatic with Steve Zissou* (2004) is a favorite among marine biologists, despite being a money-losing box office

flop. Bill Murray as oceanographer Steve Zissou goes on a quest to find and kill a "Jaguar Shark" that ate his best friend. Zissou's beast is large and mythical, having the teeth of a White Shark, six gill slits (sixgill shark, *Hexanchus*?), three dorsal fins (codfish?), and a unique spotted coloration. Misadventures abound, but fortunately (warning: spoiler alert!) Zissou has a change of heart in the end. Stingrays have a cameo role in the final chase scene.

Sharks play pivotal, villainous roles in other Hollywood productions. In *The Old Man and the Sea* (1958), sharks tear apart the giant Blue Marlin (*Makaira nigricans*) that Hemingway's hero Santiago finally catches after 84 days without a fish. In *The Deep* (1977), sharks, among such standard villains as giant moray eels, add drama and opportunities for fast-paced musical scoring. The credibility of the shark scenes suffers because the movie is set in the Caribbean, whereas the shark scenes clearly involved Indo-Pacific species. (Wouldn't you hate to go the movies with us?) *Deep Blue Sea* (1999; scantily clad female researcher, spring break teenagers) featured mad scientists who genetically engineer mako sharks and turn them into cunning, deceptive, and vengeful killers that threaten the world. The sharks meet an untimely end in several impressive explosions, including a finale that *MythBusters* busted. In *Open Water* (2003), a pleasure scuba dive in the Bahamas turns tragic when a husband and wife are left behind by the boat operators because of a screwed-up head count. Sharks contribute significantly to the unhappy ending, one that probably set the scuba diving industry back a decade. Two James Bond films, *Thunderball* (1965) and *The Spy Who Loved Me* (1977), gave sharks minor roles as disposal systems for Bond's legion of enemies.

Some of the best footage of free divers interacting with White Sharks, including intimate underwater scenes of divers taking advantage of tonic immobility (see "What is 'tonic immobility'?" in chapter 4), occur in *Dark Tide* (2012). The film is largely accurate biology-wise, except when Halle Berry's character informs Ralph Brown's character that you can tell a shark is a male because "I can see his claspers on his anal fins" (hint: reread "How do sharks reproduce?" in chapter 6). If anyone stays for the credits, there's a shark conservation message tacked on at the very end. This one went quickly to DVD.

On a more serious side is *Soul Surfer* (2011). The films retells the story of champion surfer Bethany Hamilton, who was attacked by what is thought to have been a 4.3-m (14-ft) Tiger Shark when she was 13 (see "Can scientists estimate the size and species of a shark from bite marks?" in chapter 9). Ms. Hamilton overcame the adversity of losing her left arm and went on to be a successful professional surfer.

Some memorable lines in movies have involved sharks: "Beneath this glassy surface, a world of gliding monsters!" (*Deep Blue Sea*), and of course,

"You're gonna need a bigger boat" (*Jaws*). The *Jaws* theme may be the best-known two-note musical phrase in the world.

Hollywood also puts sharks in the title of movies even when nothing resembling an elasmobranch ever appears. Hence, we have *Swimming with Sharks*, *Sharky's Machine*, *Park Sharks*, *Steel Sharks*, *Shark*, *Eagle vs. Shark* (although any shark nut would love to have the Halloween costume the actress wore), *The Adventures of Sharkboy and Lavagirl*, *Shark City*, and more.

DOCUMENTARIES. *Sharkwater* (2006) is an independently produced film about sharks, their behavior, and the problems they face. More an advocacy piece than a documentary (think "Michael Moore meets sharks"), *Sharkwater* presents an impassioned argument against the practice of finning (see "What is finning?" in chapter 10). The film has won numerous awards and raised awareness about shark conservation issues. Rent the DVD and draw your own conclusions about its accuracy now that you've read this book.

ANIMATED MOVIES. Animated films have also featured sharks, again usually in heart-pounding chase scenes that wind up with sharks as frustrated predators. *The Little Mermaid* contains one such scene. *Finding Nemo*, unquestionably the best animated film about fishes ever made, has another. *Shark Tale* gave us kinder, gentler sharks, although some of the *Shark Tale* sharks suffered personality disorders and were in obvious need of therapy to deal with their aggressive tendencies. Targeting a younger audience is the *Kenny the Shark* video series ("You'll Wish You Had a Pet Shark Too").

TELEVISION. TV documentaries about sharks abound, too many to list. These usually accurate although too often hyped depictions of nature include specials from the BBC (the *Planet Earth* and *Blue Planet* series in particular), PBS programs like *NOVA* and *Nature*, and a variety of programs on National Geographic, IMAX, Disney, Animal Planet, and the Discovery Channel, to name just a few. *Shark Week*, broadcast on the Discovery Channel each summer (just as we're all thinking it's safe to go back in the water) gets mixed reactions from shark biologists, since its immensely popular broadcasts seem designed to frighten as much as enlighten. Regardless, *Shark Week* celebrated its 25th anniversary in 2012 and draws over 30 million unique viewers in the more than 70 countries to which it is broadcast.

MonsterQuest, shown on the History Channel from 2008 to 2010 with occasional reruns, didn't really bother to enlighten viewers in its uncomfortably misinformative "Mega Jaws" (2009), purportedly about a 60-foot Megatooth Shark that was alive and well off Baja California, Mexico, despite having gone extinct 1.6 million years ago. "Monster Sharks" (2010)

speculated about the apparent recent increase in attacks by White Sharks. "Jaws in Illinois" (2008) had less material to work with—the landing of a single 1.5-m (5-ft) Bull Shark near St. Louis in 1937, an event which invited even more than the usual *MonsterQuest* speculation. Animal Planet aired a fictional documentary, *Mermaids: The Body Found* (2012), in which mermaids were depicted as having co-occurred with Megs in polar seas 1.6 million years ago. This latitude extends the known range of Meg northward; where mermaids lived is anybody's guess.

Saturday Night Live featured numerous shark spoofs about a "Land Shark" ("the cleverest of all sharks") beginning in 1975 and coinciding with the release of the movie *Jaws*. Usually voiced by Chevy Chase wearing a shark headdress, the Land Shark would knock on doors pretending to be a repairman, a flower deliveryman, a door-to-door salesman, and so on. When the door was opened, he would pounce on and devour the occupant. The Land Shark appeared in at least eight later episodes.

The presumed curative powers of shark products even made it into episodes of the TV drama *Friday Night Lights*. In season 2, Jason Street, the paralyzed high school quarterback played by Scott Porter, goes to Mexico to undergo an experimental medical procedure banned in the United States. The admittedly risky operation involves injections of "shark blood" that will allow him to walk again.

Books. There are just too many shark books around to start listing them. Good sources can be found in appendix C (websites that provide useful and accurate information on sharks); see especially the annotated book lists at www.elasmo-research.org and Elasmodiver.Com. Among the many recent engaging nonfiction books about sharks is *The Devil's Teeth* by Susan Casey. The book focuses on the White Sharks that come annually to the Farallon Islands off San Francisco to feed on sea lions and elephant seals, and on the people who conducted research there (as well as the camp followers of the researchers). Among children's literature is the popular Magic School Bus series, with some volumes about the sea containing shark encounters. Our favorite is *The Great Shark Escape*, in which Ms. Frizzle's class trip winds up in an undersea adventure. Another shark-themed series is the Shark Wars books by E. J. Altbacker. These books recount the adventures of the Shark Clans as they fight to survive in a changing ocean world.

Comic Strips. A deservedly popular cartoon strip that appears in over 250 newspapers in North America and more than 30 foreign countries is *Sherman's Lagoon*. The strip, drawn by Jim Toomey and launched in 1991, stars an overweight, rather dim-witted White Shark named Sherman who

eats just about anything but truly enjoys "hairless beach apes" (and poodles, "the other white meat"). The strip is peopled (animaled?) with a bewildering array of sea critters, many with quirky, forceful personalities. The humor in the strip can be offbeat, but we find it extremely entertaining. The action takes place chiefly in Kapupu Lagoon, west of the Rock Islands of Palau, Western Caroline Islands, one of our favorite places to dive and watch sharks (although White Sharks are admittedly rare in this part of the tropical Pacific). Earlier, Gary Larson's *The Far Side* set the standard for elasmotoons, including one of our favorites, entitled "The Heartbreak of Remoras," showing a shark bearing several attached hitchhikers.

MUSIC. Probably the best-known musical composition involving sharks (at least metaphorically) is the song "Mack the Knife." Originally part of a German production, "Mack the Knife" became the opening number in the popular musical *The Threepenny Opera*. Mack (MacHeath in the opera) is a robber and murderer whose knife attacks are likened to those of a shark: "When that shark bites, with his teeth, babe / Scarlet billows start to spread."

Translated and reworded repeatedly, the most popular version of "Mack the Knife" was recorded by Bobby Darin in 1959, but it has also been recorded or performed by Louis Armstrong, Frank Sinatra, Ella Fitzgerald, Tony Bennett, Kevin Spacey, Bill Haley & His Comets, Tito Puente, The Doors, and Sting. Darin's version earned him a Grammy Award, was number 3 on *Billboard*'s "All Time Top 100," and was ranked as number 251 on *Rolling Stone* magazine's "500 Greatest Songs of All Time." Although the BBC quickly censored it for air play because of its "lyrics about knives, dead bodies and blood," tastes changed, and it was performed at the Diamond Jubilee Concert for Queen Elizabeth II in 2012. On the other side of the musical spectrum, Frank Zappa's "Mud Shark" is a cult classic.

A few popular bands have had sharky names—for example, the bands Great White, and Sharks. Great White plays hard rock and heavy metal songs. Formed in 1977 and featuring a shifting cast of musicians, Great White is probably best known outside its musical genre for having been onstage during the tragic The Station nightclub fire in Rhode Island in which 100 people, including one band member, died. Sharks is an English punk rock band that formed in 2007 and had recorded four albums as of 2012.

VIDEO GAMES. Many video games, designed for various formats and platforms, have had shark themes. Examples are *Jaws: Ultimate Predator*, *Jaws Unleashed*, *Jaws Revenge*, *Hungry Shark*, *Shark or Die*, *Paranormal Shark Activity*, *Shark Attack 2*, *Far Cry 3—Shark Hunting + Giant Manta*

Rays, and more. Most involve shooting the sharks with a variety of weapons, a less-than-admirable goal. We can't claim expertise in any of these.

Miscellaneous Cultural References. The English language has sayings (like "swimming in shark-infested waters"), jokes ("Question: Why don't sharks attack lawyers? Answer: Professional courtesy"), and characterizations that invoke sharks. "Card shark," "pool shark," and "loan shark" all imply someone with nasty, predatory habits. Ketchikan, Alaska, artist Ray Troll turned this characterization on its head with his rendition of C. M. Coolidge's famous early-twentieth-century series of paintings, *Dogs Playing Poker*. Troll's interpretation, called *Card Sharks*, depicts a table surrounded by sharks in what appears to be a friendly game.

People who dive for a living—researchers; spearfishers; pearl, sea urchin, and abalone divers—would just as soon not deal with sharks. They have a number of ways of referring to these unwanted visitors. Research divers refer to sharks, especially aggressive offshore species such as the Galapagos Shark, as "the men in the grey suits." An elder Palauan spearfisherman once told one of us (GH), "Sharks are like police. They only show up when you least want them."

Manta Rays and Chimaeras in Popular Culture. Sharks aren't the only elasmobranchs to receive attention in popular culture. Revered in many other cultures, manta rays appear occasionally in Western culture in various forms. The SeaWorld parks in Orlando and San Diego have a roller coaster ride named "Manta," in which a model of a manta takes riders through twists, turns, and barrel rolls. In the 1970s the German carmaker Opel produced a sports car called the Manta, whose front end was supposed to at least recall the image of a manta's face (??). The iconic American sports car of our childhood was the Corvette Stingray.

Musically, an American alt-rock band named Pixies released songs titled "Manta Ray" and "Dancing the Manta Ray." Another alt-rock group, Ludo, recorded "Manta Rays" in 2010. Hollywood got in on the manta-as-monster action in the 1930s with two movies, *The Sea Bat* (1930), starring Boris Karloff before he became Frankenstein's monster, and *The Sea Fiend* (1936), later rereleased as *Devil Monster* (1946).

Among ichthyologists, a popular band that sings about chondrichthyan fishes is The Ratfish Wranglers, out of Ketchikan, Alaska. Their leader, artist Ray Troll, had a newly discovered chimaera named for him in appreciation of his efforts in educating people about sharks and their relatives. The Pointy-Nosed Blue Chimaera is unique in having a lavender color and a longer-than-average nose. Troll's nose is not much longer than average.

SHARK HOAXES. No discussion of sharks in popular culture is complete without some mention of the multitude of hoaxes that appear in magazines and, more recently, on websites. These usually involve some reference to sharks eating people or sightings of sharks in unusual places, usually involving large (and usually White) sharks. In 1996 an American tabloid called *The Sun* published an eyewitness account of giant, 75-foot-long sharks with 10-foot-wide jaws swallowing whole fishing boats in the North Atlantic. The article purported to interview scientists who speculated that the giant sharks had been frozen in glaciers that had recently melted (perhaps foretelling global warming?). The U.S. and Canadian navies supposedly launched submarines to hunt down and kill these ship-devouring monsters. Trust us: it's not true.

Many hoaxes involve fairly skillful manipulations using computer software such as Photoshop, with copyrighted images stolen from other websites and pasted into other photos. One widely distributed photo titled variously "A bad day at work" and "Picture of the Year" showed a White Shark leaping from the water to attack a South African frogman who was being hoisted up a ladder to a waiting helicopter; a large orange bridge is in the background. The helicopter is in fact in front of the Golden Gate Bridge, San Francisco, turned 180 degrees to disguise it. The shark is from the now-famous *Air Jaws* series about the White Sharks near Cape Town, South Africa, that charge out of the water in "Polaris breaches" to eat Cape fur seals (see the discussion of ambushers under "How do sharks catch their prey?" in chapter 7).

Another photo, purportedly taken from a police vehicle after record flooding in New York State in the fall of 2011, showed a White Shark swimming on the Rockaway Parkway. The shark is cropped from a well-known photo of a White Shark following a researcher in a kayak, taken by Thomas Peschak, again near Cape Town. The spectacular original color poster, along with Peschak's many stunning underwater photos, can be found at www.thomaspeschak.com.

What roles have sharks played in art and literature?

Skates more than sharks and rays appeared regularly in seventeenth- and eighteenth-century European paintings. Most depict fishmongers' stalls, draped with an abundance of fishes and shellfishes—cod, salmon, sturgeon, and flatfish predominating—plus one skate. Others are still lifes, with fishes as the main subject. Interestingly, skates are almost always shown belly up, and male skates with obvious claspers are almost as dominant. This may reflect an artistic device, or perhaps only one painter actually saw a skate, painted it that way, and later artists just copied the image.

Examples can be found in the works of Adriaen van Utrecht, Frans Snyders, Joachim Beuckelaer, Abraham van Beijeren, Giacomo Francesco Cipper, and Jean-Baptiste Siméon Chardin.

Sharks appear in more recent art work. One of the best-known paintings depicting large sharks is Winslow Homer's *The Gulf Stream* (1899). An unfortunate Caribbean fisherman lies on a small, rudderless, dismasted vessel in stormy seas, complete with a nearby waterspout. Large sharks, mouths agape, swim near the boat as a large vessel sails past in the distance.

Almost as famous, although less anatomically correct, is *Watson and the Shark* (1778) by John Singleton Copley. The dramatic painting recreates an actual attack that occurred on Brook Watson in Havana Harbor. Watson lost a leg in the attack. The painting shows a naked Watson on his back in the water about to be attacked by a large shark, as men in a small boat reach toward him unsuccessfully and another man prepares to spear the shark. The painting suggests that Copley knew little about shark anatomy: his shark has apparent lips, forward-facing eyes, and obvious rays in its fins. The painting now hangs in the National Gallery of Art, Washington, DC.

Among contemporary pieces is a well-known (some say notorious) 1991 installation by famed British artist Damien Hirst. Titled *The Physical Impossibility of Death in the Mind of Someone Living*, the work consists of a 4.3-m (14-ft) Tiger Shark, mouth agape, suspended by monofilament in a vitrine (glass tank) filled with formalin. (See http://en.wikipedia.org/wiki/The _Physical_Impossibility_of_Death_in_the_Mind_of_Someone_Living.) It is said to have sold in 2004 for $8 million. Poor preservation techniques led to decomposition of the shark, which was replaced in 2006. The piece was then loaned to the Metropolitan Museum of Art in New York, where it was displayed from 2007 to 2010.

Shark finning, or rather the campaign against finning, has inspired much artistic effort in recent years. In 2012, a traveling exhibition was sponsored by PangeaSeed, an international nonprofit shark and ocean conservation organization (www.pangeaseed.org). The exhibit focused on finning, shark conservation, and related issues. Titled *The Great West Coast Migration— Art Benefit for Sharks & Oceans*, this exhibit was shown in a half-dozen U.S. cities and featured paintings, sculptures, films, and the like, all devoted to sharks. The plight of the oceans in general has become a popular topic for contemporary artists, who put their skills to work to support many conservation causes.

Numerous "picture books" about sharks exist, but few stand out because of their artistic quality. Ray Troll's *Sharkabet* features excellent drawings of living and prehistoric species, including rays and chimaeras, and attempts to depict their likely ecological settings. Richard Ellis's *The Book of Sharks* also features accurate drawings that have an underwater feel.

Visitors to the San Francisco stopover of PangeaSeed's *Art Benefit for Sharks & Oceans* admire paintings and sculptures promoting shark conservation. Photo by Gene Helfman

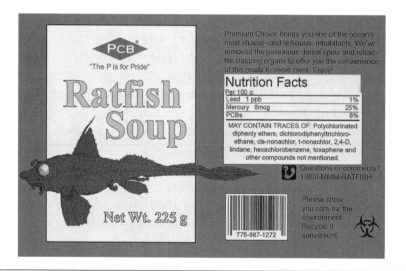

Yummm, ratfish! Part of a whimsical soup can label titled "Ratfish Soup" by Vancouver, B.C., artist Cam MacDonald (macdonaldcam@gmail.com). MacDonald included this in a 1,700-can exhibition that included can labels for several other poorly known, noncommercial fish species (diced coelacanth, viperfish fillets, barreleye fillets, crested melamphid, whole tidepool snailfish). MacDonald's work calls attention to overfishing, rampant consumerism, and the pervasiveness of pollutants in fishes that live far from human sources of toxic waste. Photo by Gene Helfman

Sharks have served as subject matter for innumerable poets. In honor of the Discovery Channel's "shark week," the Academy of American Poets publishes on its website *Poems for Shark Week*, with links to 26 shark poems, including work by Carl Sandburg, Robert Graves, Rudyard Kipling, James Dickey, and Walt Whitman (see www.poets.org/viewmedia.php/prm MID/20916). The discussion at the site notes that "sharks have long served as a cultural symbol of mortality and looming danger." Examples include Charles Wharton Stork's "Flying Fish: An Ode," in which the poet describes what it might be like to live in fear of sharks:

Enclosed above, beneath, before, behind
In green uncertainty, from which a shark
At any time may dash
And doom you like some huge demonic fate.

Other small prey fish, such as schooling menhaden, lead a similar, danger-filled existence in Isaac McLellan's "The Bluefish":

The vast menhaden multitudes
 They massacre o'er the flood;
With lashing tail, with snapping teeth
 They stain the tides with blood.

Herman Melville, author of *Moby Dick*, wrote about the relationship between pilot fish and sharks in "The Maldive Shark." Melville considered sharks "lethargic and dull" scavengers. He was much more complimentary towards the pilot fish, which he mistakenly thought sought refuge inside the shark's mouth and led sharks to food but then didn't feed on the leftovers:

The sleek little pilot-fish, azure and slim,
How alert in attendance be.
From his saw-pit of mouth, from his charnel of maw
They have nothing of harm to dread,
But liquidly glide on his ghastly flank
Or before his Gorgonian head;
Or lurk in the port of serrated teeth
In white triple tiers of glittering gates,
And there find a haven when peril's abroad,
An asylum in jaws of the Fates!
They are friends; and friendly they guide him to prey,
Yet never partake of the treat—
Eyes and brains to the dotard lethargic and dull,
Pale ravener of horrible meat.

Famous novels that feature sharks include Hemingway's *The Old Man and the Sea*. Sharks tear apart the giant Blue Marlin that Santiago finally catches. Santiago has nothing but disdain for many of the sharks that attack his fish but is highly respectful of the superpredatory mako that does the most damage. Hemingway was an avid big-game fisher and included sharks in other books, most notably *Islands in the Stream*.

Other shark-focused novels include Steve Alten's *MEG: A Novel of Deep Terror* (1996) and its sequels. In *MEG*, a deep-sea diver encounters the presumably extinct giant Megatooth Shark ("Meg" for short) in the very deep Marianas Trench (see "What is the largest shark that ever lived?" in chapter 1). A series of unfortunate events brings Meg up to shallow water, where it wreaks havoc on humanity before the hero manages to kill it, from the inside. The Alten books have not received critical acclaim.

Resurrecting Meg seems too tempting a plot device to have been ignored by other authors. In *Extinct* by Charles Wilson (1997), a nuclear test in the South Pacific awakens and frees a Meg from the Marianas Trench, again. Another Meg search sets the stage for *The Midnight Sea* by B. Luciano Barsuglia (1997). And another escaped Meg stars in *The Arc* by Paul A. Rudd (2010).

We are aware of two fairly recent graphic novels about sharks. *The Shark King*, by R. Kikuo Johnson (2012), is a retelling of the Hawaiian legend of the king of sharks (described above in "What roles do sharks play in religion and mythology?"). Matt Dembicki's *Xoc: The Journey of a Great White* (2012) follows the adventures of a Great White that travels from the Farallon Islands off San Francisco to Hawaii. The title gets its name from the Mayan word for shark, *xoc*, which is thought by some to be the origin of the English word *shark* (see "How did sharks get their name?" in chapter 1).

And of course there is *Jaws*, the novel by Peter Benchley, and its sequels.

Chapter 12

"Sharkology"

Who studies sharks?

Unlike *ichthyology*—the study of fishes—there is no accepted term for the study of sharks. *Condrichthyology* covers all the groups but is kind of clumsy. *Elasmobranchology* would work, but it excludes the holocephalan ratfishes (and is even clumsier). Most shark researchers think of themselves as just that, shark researchers, or as ichthyologists who focus on sharks and related fishes.

Regardless of name, the study of sharks and their relatives is a large, valid, and active branch of biology. Shark researchers determine how to tell different species of sharks apart and work out the evolutionary relationships among sharks. They study ecology, behavior, physiology, genetics, distribution, and conservation of sharks to answer many of the questions addressed in the first eleven chapters of this book. In fact, most of the information you've been reading about was at some point the subject of careful scientific study. Fishery biologists use this information plus additional information on life histories and population sizes to manage shark populations so that we can continue to harvest fishes for food and sport without endangering the survival of species.

Most shark researchers in North America are members of the American Elasmobranch Society (AES; www.elasmo.org). AES is a nonprofit scientific organization with an international membership of more than 400 individuals who work for universities, government agencies, and nongovernmental organizations, and as private individuals. AES "seeks to advance the scientific study of living and fossil sharks, skates, rays, and chimaeras, and the promotion of education, conservation, and wise utilization of natural

resources." Although anyone interested in the objectives of the society can join AES, membership requires endorsement, preferably by someone with a professional affiliation. The annual meetings of AES are lively affairs, usually held in conjunction with meetings of the American Society of Ichthyologists and Herpetologists, where active researchers present their latest findings, promote student research and shark conservation, and enjoy the fellowship of like-minded admirers of fishes with cartilaginous skeletons. Coauthor George Burgess is a former president of AES.

Professional organizations in other countries also focus on research and conservation of sharks and their relatives. New ones are created each year as interest and concern grow. Links to many can be found at the AES website.

Can anyone study sharks?

Professional status is not required to study sharks. It all depends on what level of "study" we're talking about. Both of your authors have been deeply involved in learning about (studying) sharks since we were quite young, and we truly hope that you, our readers, will move on from the information in this book to explore shark biology more deeply. Attend lectures, read actual scientific papers, write letters, get involved.

But to become a real (bona fide) shark researcher requires a considerable amount of professional training, beginning with a college degree in an appropriate field such as biology. Research opportunities for college students vary greatly from place to place, but if you're fortunate enough to be at an institution that has shark researchers on the faculty, drop in and say hello and say you would be happy to volunteer your time helping them. No shark researcher? Not a problem. Just volunteer with someone who studies any related field. Then work like the dickens to show you are reliable, tireless, and attentive. Take a class from that faculty member and do exceptionally well in it (this seems obvious, but we have learned otherwise). This person will be your passport to graduate school via a personal recommendation. (Of course you have to do well in your other classes too, including the ones you're not interested in.)

From there, you go on to graduate school and actually get to study sharks. Pursue a master's and ideally a Ph.D, join AES, and you have arrived.

Which species are best known?

Shark size, habitat type and depth, abundance, and activity levels are probably the chief determinants of whether we know much or little about a

shark species. A great deal of what we know about the anatomy, biochemistry, parasites, food habits, age and growth, and reproductive biology of sharks—in short, everything but their behavior and ecology—is acquired through dissection of dead offshore and inshore species caught commercially or by biologists. Tolerance of captivity is another important attribute of the sharks we know best. As a result, we know the most about abundant, commercially important species and small, slow-moving species from shallow nearshore habitats. Among larger species, those that do well in large public aquariums are better known, although we always have to be careful about drawing conclusions from animals whose behavior has been altered by being kept captive.

Putting things together, a list of best-known sharks would include Whale, Nurse, Sandtiger, Shortfin Mako, Porbeagle, Basking, Common and Bigeye threshers, Smallspotted Catshark, Smooth Dogfish, Leopard, Silky, Bull, Blacktip, Blacktip Reef, Dusky, Sandbar, Tiger, Lemon, several sharpnose sharks (*Rhizoprionodon* spp., especially *terraenovae*), Whitetip Reef, Bonnethead, Scalloped Hammerhead, Great Hammerhead, and both Spiny Dogfishes. The better-known rays include Atlantic Stingray, mantas, Cownose and Bat rays, Round Stingray, Thorny Skate, Barndoor Skate, Little Skate, Winter Skate, and Thornback Skate.

Which species are least known?

Flip the lists above, and you have the answer. Large, deepwater, offshore, and rare species; active species intolerant of captivity; and most freshwater species are poorly known (except for what has been learned from dissec-

Leopard Sharks are medium-sized animals (up to 1.5–1.8 m, or 4.9–5.9 ft) that frequent shallow waters from Oregon to northern Mexico. Their biology is well known. They feed on a variety of crabs, clams, and small fishes and adapt well to captivity. Photo by Matthew Field, www.photography.mattfield.com, Wikimedia Commons, http://en.wikipedia.org/wiki/File:Leopard _shark_in_kelp.jpg

The Frilled Shark. Sometimes referred to as living fossils, Frilled Sharks are rare inhabitants of moderate depths (1,000–1,500 m, or 3,300–4,900 ft) and are considered to be among the most primitive of modern elasmobranchs because of their tooth type, jaw connection to braincase, and six gill slits (placing them with the hexanchiform cow sharks). Not seen alive in their natural habitat until 2004, only a few have been filmed in the wild. The videos show that Frilled Sharks have a very eel-like swimming motion, a behavior we would expect, given their long body and short dorsal, anal, and pelvic fins set back near the tail. Wikimedia Commons, photo by Citron; http://en.wikipedia.org/wiki/File:Chlamydoselachus_anguineus_head.jpg

tions of dead sharks). So among the poorly known species are some lamnoid sharks (White, Megamouth); most carcharhiniforms; the hundreds of deep-sea sharks, including Frilled Sharks (*Chlamydoselachus anguineus*), goblin sharks, and sleeper sharks; and most skates and rays (including the highly endangered sawfishes and chimaeras). Rare, usually small elasmobranchs, too numerous to mention, would add greatly to the list.

How do scientists tell sharks apart?

Taxonomy is the science devoted to identifying, describing, and naming organisms. Taxonomists develop the identification guides that other people use to tell sharks apart.

If you look in such a guide to the species of sharks, you will often find at the outset a "comparative key to species" of a particular region, nation, or even the world, often including drawings of the parts of a shark or ray that are most useful for determining identification. Most such keys focus on external anatomy and the relative dimensions of body parts. For sharks, tooth characteristics are extremely useful, as are fin size, shape, and location (which is why we seldom get good identifications of attacking sharks; see "Are sharks dangerous to people?" in chapter 9). Body proportions (depth

Sharks: The Animal Answer Guide

relative to length of different parts), head shape, mouth width, lateral-line pore distribution (especially in chimaeras), coloration, scale roughness and distribution (especially in skates), eye diameter and color and pupil shape, number of gills, and barb characteristics (in stingrays) are also helpful. Information about geographic location is absolutely essential, but such "locale data" are sometimes lost in the case of preserved museum specimens. A researcher may also dissect the animal and look at the size, number, shape, and distribution of various internal organs. Increasingly, biochemical characteristics, including molecular genetics (i.e., DNA fingerprinting), are often used in combination with one or more of the anatomical traits to finalize an identification.

By making use of these characteristics, we can tell what species a shark may be or whether it is unknown to science. In addition, molecular techniques have been used increasingly in forensic investigations to determine if shark meat or fins being sold are in fact from the species advertised. These techniques have allowed researchers to ascertain when parts of rare, protected species are being passed off illegally as being from abundant, unprotected species.

How do scientists study the movements of large sharks?

One of the big questions in elasmobranch biology has been and remains how much sharks travel and where they go. This critical piece of information is essential for conserving sharks because many move long distances. This means protection in one place may be ineffective if the same animal moves to an area where it isn't protected.

For a long time, the only way we could gain such information was to tag a shark by placing an individually recognizable tag on its fin or in its back, and hope someone would catch (and probably kill) it somewhere else and report the tag number to the address listed on the tag, hopefully with size and capture locale information included. Moving needles and giant haystacks.

Slowly, we've learned that certain species—White Sharks, Whale Sharks, Blacktip Reef Sharks, Whitetip Reef Sharks, Zebra Sharks, manta rays and Spotted Eagle Rays—have individually distinctive markings. These markings sometimes make photo-identification possible, whether a researcher is underwater looking at color patterns, or above water in a boat looking at the dorsal fin shape and scarring of White Sharks (see "Can sharks tell one another apart?" in chapter 4).

As technologies have grown more sophisticated and complex electronic equipment has been miniaturized, researchers have developed "packages" that can be surgically implanted, attached to the dorsal fin, glued on the

shark's back, or inserted via small harpoons into back musculature. Ultrasonic tags that transmit frequency-coded signals are one such technology. At first, these methods required researchers to sit for hours in small boats, maps on their laps, listening through hydrophones (underwater receivers) and recording the location whenever they were lucky enough to hear the animal's "pinger." The data collected were usually relatively short-term because of battery life and researcher fatigue issues. Vast improvements include better, longer-lasting batteries and arrays of recording hydrophones placed on the seafloor. The hydrophones are retrieved by divers after weeks or months, and the data are then downloaded.

Most recently, advances have allowed the development of satellite tracking and "archival" tags that periodically record depth, temperature, and geological location of the tagged animals. Some tags transmit data to satellites whenever a shark comes to the surface; others eventually pop up and send out a signal that allows researchers to physically retrieve the tag and access the incredible storehouse of encoded information. Such tags have allowed us to follow intimately the long-term and long-distance movements of such animals as White, Basking, Blue, Bull, Tiger, Salmon, Shortfin Mako, and Whale sharks, and numerous smaller sharks and rays that undergo less spectacular but equally informative movements. For an example of such a program, see the website Tagging of Pacific Predators (TOPP) at www.topp.org.

Appendix A

Sharks, Skates, Rays, and Chimaeras
of the World

The following classification is based on Nelson (2006), with some updates from various experts.

Class Chondrichthyes (cartilaginous fishes)
　Subclass Elasmobranchii (sharklike fishes)
　　Infraclass Euselachii (sharks and rays)
　　　Division Neoselachii
　　　　Subdivision Selachii (sharks)
　　　　　Superorder Galeomorphi
　　　　　　Order Heterodontiformes[a] (8 species, marine):
　　　　　　　Heterodontidae[b] (bullhead, horn sharks)
　　　　　　Order Orectolobiformes (32 species, marine):
　　　　　　　Parascylliidae (collared carpet sharks)
　　　　　　　Brachaeluridae (blind sharks)
　　　　　　　Orectolobidae (wobbegongs)
　　　　　　　Hemiscylliidae (bamboo sharks)
　　　　　　　Stegostomatidae (Zebra Shark)[c]
　　　　　　　Ginglymostomatidae (nurse sharks)
　　　　　　　Rhincodontidae (Whale Shark)
　　　　　　Order Lamniformes (15 species, marine):
　　　　　　　Odontaspididae (sand tiger sharks)
　　　　　　　Mitsukurinidae (Goblin Shark)
　　　　　　　Pseudocarchariidae (Crocodile Shark)
　　　　　　　Megachasmidae (Megamouth Shark)
　　　　　　　Alopiidae (thresher sharks)
　　　　　　　Cetorhinidae (Basking Shark)
　　　　　　　Lamnidae (mackerel sharks)
　　　　　　Order Carcharhiniformes (224 species, mostly marine):
　　　　　　　Scyliorhinidae (catsharks)
　　　　　　　Proscylliidae (finback catsharks)
　　　　　　　Pseudotriakidae (false catsharks)

[a] The terms used for orders of sharks end in the letters *-iformes*. Adjective and noun forms end in *-iform* and *-iforms*. For example, the heterodontiform bullhead sharks are also called heterodontiforms.

[b] The names for families of sharks always end in the letters *-idae*. Adjective and noun forms end in *-id* and *-ids*. For example, the heterodontid bullhead sharks are also called heterodontids.

[c] When a family contains only one species, we have capitalized the species name, using the convention of capitalizing common names of specific species.

Leptochariidae (Barbeled Houndshark)
Triakidae (houndsharks)
Hemigaleidae (weasel sharks)
Carcharhinidae (requiem sharks)
Sphyrnidae (hammerhead sharks)
Superorder Squalomorphi
Order Hexanchiformes (5 species, marine):
Chlamydoselachidae (Frill Shark)
Hexanchidae (cow sharks)
Order Echinorhiniformes (2 species, marine):
Echinorhinidae (bramble sharks)
Order Squaliformes (97 species, marine):
Squalidae (dogfish sharks)
Centrophoridae (gulper sharks)
Etmopteridae (lantern sharks)
Somniosidae (sleeper sharks)
Oxynotidae (rough sharks)
Dalatiidae (kitefin sharks)
Order Squatiniformes (15 species, marine):
Squatinidae (angel sharks)
Order Pristiophoriformes (5 species, marine):
Pristiophoridae (saw sharks)
Subdivision Batoidea (skates and rays)
Order Torpediniformes (59 species, marine):
Torpedinidae (torpedo electric rays)
Narcinidae (numbfishes)
Order Pristiformes (7 species, marine and freshwater):
Pristidae (sawfishes)
Order Rajiformes (285 species, marine):
Rhinidae (bowmouth guitarfishes)
Rhynchobatidae (wedgefishes)
Rhinobatidae (guitarfishes)
Rajidae (skates)
Order Myliobatiformes (183 species, marine and freshwater):
Platyrhinidae (thornbacks)
Zanobatidae (Striped Panray)
Hexatrygonidae (Sixgill Stingray)
Plesiobatidae (Deepwater Stingray)
Urolophidae (round stingrays)
Urotrygonidae (American round stingrays)
Dasyatidae (whiptail stingrays)
Potamotrygonidae (river stingrays)
Gymnuridae (butterfly rays)
Rhinopteridae (cownose rays)
Mobulidae (manta rays, mobulas)
Subclass Holocephali
Superorder Holocephalimorpha (6 extinct orders and modern
chimaeras)

Order Chimaeriformes (modern chimaeras) (46 species, marine):

Callorhinchidae (plownose chimaeras)

Rhinochimaeridae (longnose chimaeras)

Chimaeridae (shortnose chimaeras or ratfishes)

Appendix B

Organizations That Promote the Study and Conservation of Sharks

American Elasmobranch Society. Focuses on biology and conservation of sharks, skates, and rays. See www.elasmo.org.

American Fisheries Society. Publisher of *Transactions of the American Fisheries Society*, *North American Journal of Fisheries Management*, *North American Journal of Aquaculture*, *Journal of Aquatic Animal Health*, and the newsletter *Fisheries*. Their website is fisheries.org.

American Institute of Fishery Research Biologists. Publishes the newsletter *Briefs*. See www.aifrb.org.

American Society of Ichthyologists and Herpetologists. Publisher of the journal *Copeia*. The website is asih.org.

Canadian Shark Research Laboratory, Bedford Institute of Oceanography, Dartmouth, Nova Scotia, Canada. Provides information about shark biology and fisheries on its website, www.marinebiodiversity.ca/shark.

Defenders of Wildlife. Maintains a website that contains information about sharks and works to disseminate information about finning. See www .defenders.org/sharks/sharks and www.defenders.org/sharks/defenders -action-stop-shark-finning.

European Elasmobranch Association (EEA). Nonprofit organization that coordinates the activities of European organizations dedicated to the study, management, or conservation of sharks, skates, rays, and chimaeras (www.eulasmo.org).

European Ichthyological Society. Sponsor of the European Congresses of Ichthyology.

Fishery Society of the British Isles (www.fsbi.org.uk). Publisher of the *Journal of Fish Biology*.

Fundación Colombiana para la investigación y conservación de tiburones y rayas (www.squalus.org). Colombian-based scientific organization that promotes the study and conservation of sharks and rays.

Ichthyological Society of Japan. Publisher of two journals, one in English, *Ichthyological Research*, and one in Japanese, *Japanese Journal of Ichthyology*.

International Union for the Conservation of Nature (IUCN). Publishes the Red List of Threatened Species (www.redlist.org) with periodic updates on the global conservation status of species.

Manta Trust. Focuses on the worldwide conservation of manta rays and their habitats (www.mantatrust.org).

Oceania Chondrichthyan Society (OCS). Australian-based organization that promotes education, conservation and scientific study of chondrichthyan fishes; see its website at www.oceaniasharks.org.au.

Pew Environmental Initiatives, Global Shark Conservation Campaign. The campaign works to influence fishing nations and treaty organizations, declare shark sanctuaries, and advocate for international shark conservation (www.pewenvironment.org/campaigns).

Project AWARE. International organization of scuba divers that promotes underwater conservation, focusing particularly on sharks and marine debris (www.projectaware.org).

Sea Shepherd Conservation Society. International organization that works to defend marine wildlife, including sharks. Its shark page and links begin at www.seashepherd.org/sharks.

Shark Advocates International. "Safeguarding sharks through sound conservation" (www.sharkadvocates.org).

Shark Alliance. Global, not-for-profit coalition of nongovernmental organizations dedicated to restoring and conserving shark populations by improving shark conservation policies (www.sharkalliance.org).

Shark Foundation. Swiss-based conservation organization (www.shark.ch).

Appendix C

Websites That Provide Useful and Accurate Information on Sharks

The websites listed below are more general than the sites listed in appendix B, which are maintained primarily by professional scientific and conservation organizations.

A special note: Beware the Internet! There are dozens, maybe hundreds, of shark websites. But the Web is often a merry-go-round and sometimes a minefield for anyone looking for reliable information about sharks. The vast majority of sites contain information taken from other websites, with little or no effort given to verifying accuracy. We have often found a train of sites containing similar information that turn out to be circular: Site A gets info from Site B, which got it from C, which got it from A, and not one of these sites says where the facts actually came from. Some sites are essentially personal web pages or blogs, a sort of mysharksite.com, and others appear to be little more than lures for advertising dollars. An example of the latter is Ask.com (which is especially sloppy when it comes to providing accurate information). The key to separating the wannabes from the helpful sites is whether they include the sources of their information. Listed below are sites that we have found particularly useful.

Arkive.com. Good for photos and videos plus some biological information.

Australian Museum (http://australianmuseum.net.au). Very good species accounts.

Elasmodiver.com. This site has a field guide to sharks and rays with many good photos, plus an extensive, annotated list of coffee-table books, shark biology books, field guides, individual species accounts, magazines, etc., at www.elasmodiver.com/Shark books.htm.

Ichthyology at the Florida Museum of Natural History (FLMNH) (www .flmnh.ufl.edu/fish/sharks/sharks.htm). Everything sharky: invaluable species accounts, conservation news, research summaries, and updated information on shark attacks worldwide.

ReefQuest Centre for Shark Research (www.elasmo-research.org). See especially the "Biology of Sharks & Rays" online course and the wonderfully detailed list of "Recommended Shark Books."

Shark Research Committee (www.sharkresearchcommittee.com). The focus is on documenting shark attacks from the Pacific Coast of North America and conducting research on the biology of Pacific Coast sharks. The organization maintains a newsletter, *Pacific Coast Shark News*.

Shark Research Institute (Australia) (www.shark.org.au). Gathers data for the global shark attack file and maintains an online database on attacks in Australia and surrounding waters.

Shark Research Institute (U.S.) (www.sharks.org). A nonprofit international scientific organization dedicated to shark conservation.

Shark Trust (www.sharktrust.org). A United Kingdom–based organization that aims to advance worldwide conservation of sharks through science, education, influence and action; publishes a newsletter, *Shark Focus*.

Shark Year Magazine (http://sharkyear.com). An online magazine updated regularly with news articles plus access to the scientific literature.

South African White Shark Research Institute (www.whiteshark.co.za). A nonprofit dedicated to research on the White Shark and preservation of its environment.

U.N. Food and Agriculture Organization (FAO). Useful fact sheets on shark species at www.fao.org/fishery/species/search/en.

Wikipedia. Much information, especially in the entries that focus on individual species.

Bibliography

Hundreds and hundreds of books have been written about sharks and their biology, fisheries, and conservation. Some books are primarily personal accounts of experiences with sharks by researchers, journalists, adventurers, and fishers. Many could be considered coffee-table books because of their emphasis on photographs; fortunately, such photos are becoming increasingly available on the web. The following is a selective catalog of books and review articles that we have found especially useful, particularly those with a more scientific focus.

Barton, M. 2006. *Bond's Biology of Fishes*. 3rd ed. Stamford, CT: Thomson Brooks/Cole.

Bigelow, H. B., and W. C. Schroeder. 1948. Sharks. In J. Tee-Van, C. M. Breder, A. E. Parr, W. C. Schroeder, and L. P. Schultz, eds., *Lancelets, Cyclostomes, Sharks*. Part 1 of *Fishes of the Western North Atlantic*. Memoirs of the Sears Foundation for Marine Research. New Haven, CT: Yale University.

———. 1953a. Chimaeroids. In H. B. Bigelow and W. C. Schroeder, eds., *Sawfishes, Guitarfishes, Skates, Rays, and Chimaeroids*. No. 1, Part 2 of *Fishes of the Western North Atlantic*. Memoirs of the Sears Foundation for Marine Research. New Haven, CT: Yale University.

———. 1953b. Batoidae. In H. B. Bigelow and W. C. Schroeder, eds., *Sawfishes, Guitarfishes, Skates, Rays, and Chimaeroids*. No. 2, Part 1 of *Fishes of the Western North Atlantic*. Memoirs of the Sears Foundation for Marine Research. New Haven, CT: Yale University.

Branstetter, S., ed. 1993. *Conservation Biology of Elasmobranchs*. NOAA Technical Report NMFS 115. Springfield, NJ: U.S. Department of Commerce.

Budker, P. 1971. *The Life of Sharks*. New York: Columbia University Press.

Camhi, M., S. Fowler, J. Musick, A. Brautigan, and S. Fordham. 1998. *Sharks and Their Relatives: Ecology and Conservation*. IUCN Occasional Paper No. 20. Cambridge, UK, and Gland, Switzerland: International Union for the Conservation of Nature.

Camhi, M. D., E. K. Pikitch, and E. A. Babcock, eds. 2008. *Sharks of the Open Ocean: Biology, Fisheries, and Conservation*. Oxford, UK: Blackwell Publishing.

Carrier, J. C., J. A. Musick, and M. R. Heithaus, eds. 2004. *Biology of Sharks and Their Relatives*. Boca Raton, FL: CRC Press.

———, eds. 2010. *Sharks and Their Relatives, II: Biodiversity, Adaptive Physiology, and Conservation*. Boca Raton, FL: CRC Press.

Carwardine, M., and K. Watterson. 2002. *The Shark Watchers Handbook: A Guide to Sharks and Where to See Them*. Princeton, NJ: Princeton University Press.

Castro, J. 2011. *The Sharks of North America*. Oxford, UK: Oxford University Press.

Collette, B. B., and G. Klein-MacPhee, eds. 2002. *Bigelow and Schroeder's Fishes of the Gulf of Maine*. 3rd ed. Washington, DC: Smithsonian Institution Press.

Collier, R. S. 2003. *Shark Attacks of the Twentieth Century: From the Pacific Coast of North America*. Chatsworth, CA: Scientia Publishing.

Compagno, L. J. V. 1984a. *Sharks of the World: An Annotated and Illustrated Catalogue of Shark Species Known to Date*, Part 1, *Hexanchiformes to Lamniformes*. FAO Fisheries Synopsis 125, vol. 4, pt. 1. pp. 1–249.

———. 1984b. *Sharks of the World*, Part 2, *Carcharhiniformes*. FAO Fisheries Synopsis 125, vol. 4, pt. 2. pp. 251–655.

———. 2001. *Sharks of the World*, Vol. 2, *Bullhead, Mackerel, and Carpet Sharks (Heterodontiformes, Lamniformes, and Orectolobiformes)*. FAO Species Catalogue for Fishery Purposes, no. 1, vol. 2.

Compagno, L. J. V., M. Dando, and S. Fowler. 2005. *Sharks of the World*. Princeton, NJ: Princeton University Press.

Compagno, L. J. V., D. A. Ebert, and S. Fowler. 2013. *Sharks of the World: A Fully Illustrated Guide*. Plymouth, UK: Wild Nature Press.

Domeier, M. L., ed. 2012. *Global Perspectives on the Biology and Life History of the Great White Shark*. Boca Raton, FL: CRC Press.

Ebert, D. A., and J. Sulikowski, eds. 2009. *Biology of Skates*. Developments in Environmental Biology of Fishes No. 27. New York: Springer.

Ellis, R. 1976. *The Book of Sharks*. New York: Grosset and Dunlop.

Ellis, R. and J. E. McCosker. 1991. *Great White Shark*. New York: Harper Collins.

Fowler, S. L., et al. 2005. *Sharks, Rays, and Chimaeras: The Status of the Chondrichthyan Fishes*. IUCN/SSC Shark Specialist Group. Gland, Switzerland: World Conservation Union.

Galucci, V., G. McFarlane, and G. Bargmann, eds. 2009. *Biology and Management of Dogfish Sharks*. Bethesda, MD: American Fisheries Society.

Gruber, S. H., ed. 1990. *Discovering Sharks. Underwater Naturalist* 19, no. 4; 20, no. 1, Special Publication 14, Bulletin of the American Littoral Society.

Hamlett, W. C., ed. 1999. *Sharks, Skates, and Rays: The Biology of Elasmobranch Fishes*. Baltimore, MD: Johns Hopkins University Press.

———, ed. 2005. *Reproductive Biology and Phylogeny of Chondrichthyes: Sharks, Batoids, and Chimaeras*. Enfield, NH: Science Publishers.

Helfman, G. S. 2007. *Fish Conservation: A Guide to Understanding and Restoring Global Aquatic Biodiversity and Fishery Resources*. Washington, DC: Island Press.

Helfman, G. S., D. E. Facey, B. B. Collette, and B. E. Bowen. 2009. *The Diversity of Fishes: Biology, Evolution, and Ecology*. 2nd ed. Hoboken, NJ: Wiley-Blackwell.

Hodgson, E. S., and R. F. Mathewson, eds. 1978. *Sensory Biology of Sharks, Skates, and Rays*. Arlington, VA: Office of Naval Research.

Klimley, A. P. 2003. *The Secret Life of Sharks*. New York: Simon and Schuster.

———. 2013. *The Biology of Sharks and Rays*. Chicago: University of Chicago Press.

Klimley, A. P., and D. G. Ainley, eds. 1996. *Great White Sharks: The Biology of Carcharodon carcharias*. San Diego, CA: Academic Press.

Kyne, P. M., and C. A. Simpfendorfer. 2007. *A Collation and Summarization of Available Data on Deepwater Chondrichthyans: Biodiversity, Life History, and Fisheries*. IUCN/SSC Shark Specialist Group. Gland, Switzerland: World Conservation Union.

Lineaweaver, T. H., and R. H. Backus. 1970. *The Natural History of Sharks.* New York: Lippincott.

Long, J. A. 1995. *The Rise of Fishes.* Baltimore, MD: Johns Hopkins University Press.

Maisey, J. G. 1996. *Discovering Fossil Fishes.* New York: Henry Holt.

Moss, S. A. 1984. *Sharks: A Guide for the Amateur Naturalist.* Englewood Cliffs, NJ: Prentice-Hall.

Moyle, P. B., and J. J. Cech. 2004. *Fishes: An Introduction to Ichthyology.* 5th ed. Upper Saddle River, NJ: Prentice-Hall.

Musick, J. A., ed. 1999. *Life in the Slow Lane: Ecology and Conservation of Long-Lived Marine Animals.* American Fisheries Society Symposium 23. Bethesda, MD: American Fisheries Society.

Nelson, J. S., chair. 2004. *Common and Scientific Names of Fishes from the United States, Canada, and Mexico.* 6th ed. American Fisheries Society Special Publication 29. Bethesda, MD: American Fisheries Society.

Nelson, J. S. 2006. *Fishes of the World.* 4th ed. Hoboken, NJ: John Wiley.

Paxton, J. R., and W. N. Eschmeyer, eds. 1998. *Encyclopedia of Fishes.* 2nd ed. San Diego, CA: Academic Press.

Peschak, T., and M. C. Scholl. 2007. *South Africa's Great White Shark.* Cape Town, South Africa: Random House Struik.

Pratt, H. L. Jr., S. H. Gruber, and T. Taniuchi, eds. 1990. *Elasmobranchs as Living Resources: Advances in the Biology, Ecology, Systematics, and the Status of Fisheries.* NOAA Technical Report NMFS 90. NOAA/National Marine Fisheries Service.

Schulze-Haugen, M., T. Corey, and N. E. Kohler. 2004. *Guide to Sharks, Tunas, and Billfishes of the U.S. Atlantic and Gulf of Mexico.* Silver Spring, MD: NOAA and Rhode Island Sea Grant.

Skomal, G. 2008. *The Shark Handbook: The Essential Guide for Understanding the Sharks of the World.* Kennebunkport, ME: Cider Mill Press.

Smith, M., D. Warmolts, D. Thoney, and R. Hueter, eds. 2004. *The Elasmobranch Husbandry Manual: Captive Care of Sharks, Rays, and Their Relatives.* Columbus, OH: Ohio Biological Survey.

Springer, V. G., and J. P. Gold. 1989. *Sharks in Question: The Smithsonian Answer Book.* Washington, DC: Smithsonian Institution Press.

Stafford-Dietsch, J. 1988. *Shark: A Photographer's Story.* San Francisco: Sierra Club Books.

Stevens, J. D., ed. 1999. *Sharks.* New York: Check Mark Books.

Yano, K., J. F. Morrissey, Y. Yabumoto, and K. Nakaya, eds. 1997. *Biology of the Megamouth Shark.* Tokyo: Tokai University Press.

Wallace, S., and B. Gisborne. 2006. *Basking Sharks: The Slaughter of BC's Gentle Giants.* Vancouver, BC: New Star Books.

WildAid/Oceana. 2007. *The End of the Line? Global Threats to Sharks.* http://oceana .org/sites/default/files/o/fileadmin/oceana/uploads/Sharks/EndoftheLine _Spread_sm.pdf.

Index

Page numbers in *italics* indicate illustrations.

Index

underwater photography, 64
Undulate Ray (*Raja undulata*), 59

Velvet Belly Lantern Shark (*Etmopterus spinax*), 60, 63, 136, 153
Vermiculate River Stingray (*Potamotrygon castexi*), 81
vertebrae: age determination by, 132–33; fossilized, 20; hardened cartilage, 32, 33; tools and ornaments, 208
vertical movements, 45, 88–90
Vibrio infections, 180–81
video games, 214–15
vision, 36–38, 49, 60, 140

warm-blooded animals, 24–25
water temperature, 13, 23–24; global warming, 191; migration, 84–90; species diversity, 95–97
websites, 232–33
Whale Shark (*Rhincodon typus*), 1, 4, *69*, *150*, 223, 227; albino, 61; avoiding predators, 105; boat injuries, 200; buoyancy, 45; coloration, 56, 59; conservation, 188, 195; enemies, 103; feeding, 15, 28–29, 68, 69, 138, 148–50; fishing, 188, 194, 195; groups, 68, 69; identifying individuals, 72; maturity and longevity, 130; migration, 84, 86, 90; remoras on, 77; reproduction, 121, 122, 123, 125; seeing in captivity, 169, *189*; seeing in the wild, 165; sex segregation, 128; size, 14–15, 18, 130; teeth, 28, 30, 105; tracking movements, 225, 226; vertical movements, 45, 89
Whitenose Shark (*Nasolamia velox*), 56
White Shark (*Carcharodon carcharias*), xvi, 1, 10, *31*, 33, *33*, *62*, *74*, *153*, 224; activity, 99, 151; albino, 61, 62; archeological artifact, 208, *209*; attacks, 172–74, 180, 216; avoiding predators, 104–5; bite force, 153; body temperature, 24; brain, 51–52, 80; buoyancy, 45; conservation, 195; in culture, 204, 209; declining numbers, 188; enemies, 27, 101, 103; evolution, 18; eyes, 36–37, 64; feeding,

23, 24, 30–31, 68, 102, 132, 136, 138, 140, 145, 151–53, 156; fighting, 73, 74, 76; fishing, 188, 194, 195; groups, 68; growth rate, 132; historical descriptions, 206; hoaxes, 216; identifying individuals, 72; intelligence, 52; maturity and longevity, 130, 131; media depictions, 210, 211, 213–14, 220; migration, 86, 87, 90; navigation, 90; olfaction, 33–34; oxygen, 23; parasites, 108; pheromones, 34; play, 82; in religion, 205; reproduction, 116, 124; safety of eating, 182; seeing in captivity, 168–69; seeing in the wild, 165–67, *166*; size, 15, 130; swimming, 43; teeth, 18, 25, 30–31, *209*; territoriality, 76; tonic immobility, 83; tracking movements, 45, 86; vertebra, 33; vertical movements, 45, 89
Whitespotted Bamboo Shark (*Chiloscyllium plagiosum*), 32, 61, 80, 114, 148
White-Spotted Marble Electric Ray (*Torpedo marmorata*), 61, 124
Whitetip Reef Shark (*Triaenodon obesus*), 223; activity, 68–69; breathing, 43; coloration, 56; feeding, 156; groups, 68–69, 71, 72; habitats, 77; lying on the bottom, 41, 43, 69; reproduction, 120; tracking movements, 225
Winghead Shark (*Eusphyra blochii*), 48, 49
Winter Skate, 88, 223
winter survival, 99–100
wobbegongs, 8, 41, 43, 54, 58, 63, 101, 105, 167
worm infestations, 108–10

Xenacanthus, 13

Yellow Smoothhound Shark (*Mustelus higmani*), 61
Yellow Stingray (*Urobatis jamaiciensis*), color vision, 38

Zambezi Shark, 173. *See also* Bull Shark
Zebra Bullhead Shark (*Heterodontus zebra*), *28*, 58, 65
Zebra Shark (*Stegostoma fasciatum*), 4, 12, 58, 62, 65, *65*, 72, 115, 127, 157, 168, 225